LONDON MATHEMATICAL SOCIETY STUDENT TEXTS

Managing editor: Dr C.M. Series, Mathematics Institute
University of Warwick, Coventry CV4 7AL, United Kingdom

London Mathematical Society Student Texts 24

Lectures on Elliptic Curves

J.W.S. Cassels
Department of Pure Mathematics and Mathematical Statistics,
University of Cambridge

CAMBRIDGE
UNIVERSITY PRESS

CAMBRIDGE UNIVERSITY PRESS
Cambridge, New York, Melbourne, Madrid, Cape Town, Singapore,
São Paulo, Delhi, Dubai, Tokyo, Mexico City

Cambridge University Press
The Edinburgh Building, Cambridge CB2 8RU, UK

Published in the United States of America by
Cambridge University Press, New York

www.cambridge.org
Information on this title: www.cambridge.org/9780521425308

© Cambridge University Press 1991

First published 1991
Reprinted 1992, 1995

A catalogue record for this publication is available from the British Library

Library of Congress Cataloguing in Publication Data

ISBN 978-0-521-41517-0 Hardback
ISBN 978-0-521-42530-8 Paperback

London Mathematical Society Student Texts 24

Lectures on Elliptic Curves

J.W.S. Cassels
Department of Pure Mathematics and Mathematical Statistics,
University of Cambridge

CAMBRIDGE
UNIVERSITY PRESS

CAMBRIDGE UNIVERSITY PRESS
Cambridge, New York, Melbourne, Madrid, Cape Town, Singapore,
São Paulo, Delhi, Dubai, Tokyo, Mexico City

Cambridge University Press
The Edinburgh Building, Cambridge CB2 8RU, UK

Published in the United States of America by
Cambridge University Press, New York

www.cambridge.org
Information on this title: www.cambridge.org/9780521425308

First published 1991
Reprinted 1992, 1995

A catalogue record for this publication is available from the British Library

Library of Congress Cataloguing in Publication Data

ISBN 978-0-521-41517-0 Hardback
ISBN 978-0-521-42530-8 Paperback

Contents

0

Introduction

Diophantine equations, that is to say equations whose solution is to be found in integers, or, alternatively, in rationals, have fascinated man from the earliest times: a Babylonian clay tablet dated to between 1600 and 1900 B.C. lists 15 solutions of the "Pythagorean" equation

$$X^2 + Y^2 = Z^2.$$

Diophantos himself lived in Alexandria in the 3rd Century A.D. We shall meet some of his ideas. His work was continued by Hypatia, the only female mathematician of antiquity whose name has come down to us. (She was cruelly done to death by the Christians: their leader was canonized.) Another mathematician whose ideas continue to play a key role is Fermat (1601-1665). For a fuller historical account in a modern context, see A. Weil *Number theory: an approach through history from Hammurabi to Legendre* (Birkhäuser, 1983). [For Hypatia, see Gibbon *Decline and Fall.*]

In this course we concentrate attention on rational solutions of Diophantine equations. The study of integral solutions requires further considerations, which we shall not touch on.

It is now clear that an appropriate language to discuss many aspects of Diophantine equations is that of algebraic geometry: not so much the classical algebraic geometry, which works over the complex numbers, but a version working over a general ground field such as the field **Q** of rationals and often called "Diophantine geometry". Some of the arguments and results of classical geometry go over to Diophantine geometry unchanged, for some the conclusions are more limited, and for others we must make further hypotheses which are automatically satisfied in the classical theory.

Diophantine equations can be interpreted as questions about the existence of points on algebraic varieties. Here we will be concerned only with curves. Geometers classify curves by a non-negative integer, the genus. The Diophantine theory of curves of genus 0 is well understood. For curves of genus 1, there is a rich body of well-established theory and an equally rich corpus of conjecture which is currently beginning to succumb to intensive research. The Diophantine theory of curves of genus > 1 is in a rudimentary state (despite Faltings' Theorem).

The main subject of this course is some of the basic Diophantine theory of curves of genus 1. To set the scene, we start with an account of genus 0. Here the situation is dominated by the local-global principle (Hasse principle). This relates behaviour over the rational field $\mathbf{Q}$ to that over its local completions, the p-adic fields $\mathbf{Q}_p$, where things are simpler. A unifying theme for curves of genus 1 is the extent to which local (i.e. p-adic) behaviour determines rational behaviour. This material generalizes smoothly to algebraic number fields but we have restricted attention to the rationals in the belief that new concepts are easiest acquired in the simplest contexts.

The final three sections mark a change of goal. Two of them introduce the more sophisticated theory over finite fields, culminating in the estimates for the number of points known as the "Riemann hypothesis for function fields" (of genus 1). The very last section indicates how these ideas are used in the modern technology for factorizing large integers.

Prerequisites.

In this course the prerequisites have been reduced to a minimum. We have spoken above about curves of genus 0 and 1, but the focus will be on concrete classes of curves such as conics and plane cubics. The p-adic numbers are introduced from scratch. A knowledge of algebraic number theory is not required, provided that the reader is prepared to take one statement on trust. Algebraic number theory is, however, indispensable for many applications, as we shall indicate in optional passages. We do require the rudiments of Galois theory: indeed one of the interests will be its application in novel contexts.

1

Curves of genus 0. Introduction

We shall say that a point is *rational*, or *defined over* **Q**, if its co-ordinates are rational. A curve is said to be *defined over* **Q** if it is given by an equation or equations with coefficients in **Q**. [Unfortunately the term "rational curve" was preempted by the geometers as a synonym for "curve of genus 0".] More generally we shall say that we are working over **Q**, or that the ground field is **Q**, if all the coefficients of the algebraic expressions involved are in **Q**.

Sometimes elementary geometric arguments continue to be valid when we work over **Q**. For example, consider a cubic curve such as

$$\mathcal{C}: \quad X^2 - Y^2 = (X - 2Y)(X^2 + Y^2),$$

which has a double point at the origin.

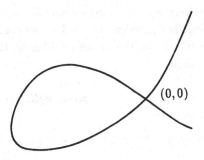

A line through the origin meets the curve in one further point, so giving

a description of all the points on the curve. More precisely, consider the
line
$$X = sY$$
for given s. This meets the curve where
$$Y^2(s^2 - 1) = Y^3(s - 2)(s^2 + 1),$$
and so in the point (x, y) where
$$x = \frac{s(s^2 - 1)}{(s - 2)(s^2 + 1)}, \qquad y = \frac{s^2 - 1}{(s - 2)(s^2 + 1)}.$$
Conversely, given (x, y) on the curve, it is of the above form with
$$s = x/y.$$
We say that C is *birationally equivalent* to the line [given by a single
variable and no equation]. In this case the birational equivalence is
defined over $\mathbf{Q}$ [i.e. the rational functions expressing the equivalence
have coefficients in $\mathbf{Q}$. Note the unfortunate clash in the double meaning
of the term "rational"].

In general there is a $1 - 1$ correspondence between the rational points
on the one curve and those on the other, the correspondence being given
by the birational correspondence. There are, however, exceptions. For
example $s = 2$ does not correspond to any point (x, y) and $s = \pm 1$ both
correspond to $(x, y) = (0, 0)$. If we had had $X^2 - 2Y^2$ instead of $X^2 - Y^2$
on the left hand side, then $(x, y) = (0, 0)$ would not correspond to any
rational value of s. It is not difficult to see, however, that if two curves
are birationally equivalent over $\mathbf{Q}$ there are only finitely many rational
points on the one which do not correspond to rational points on the
other. To study the rational points on a curve, it is thus sufficient to
consider it up to a birational equivalence defined over $\mathbf{Q}$.

A classical theorem working over the complex field $\mathbf{C}$ states that every
curve of genus 0 is birationally equivalent to the line: we could treat this
as a definition of "genus 0". When the ground field is $\mathbf{Q}$, this theorem
no longer holds. Instead we have the

Fact. *A curve of genus 0 defined over $\mathbf{Q}$ is birationally equivalent over
$\mathbf{Q}$ either to the line or to a conic.*

This reduces the Diophantine study of curves of genus 0 to that of
conics.

Theorem 1. *A conic defined over $\mathbf{Q}$ is birationally equivalent to the
line if and only if it has a rational point.*

Proof. The "only if" part is trivial. Suppose then that there is a rational point. After a change of co-ordinates we may take it to be the origin, so that the equation of the conic is

$$F_1(X,Y) + F_2(X,Y) = 0,$$

where F_j is homogeneous in X, Y of degree j. The birational equivalence with the line follows by putting $X = sY$, as in the cubic case discussed earlier.

The Diophantine theory of curves of genus 0 is thus reduced to deciding when a conic defined over **Q** has a rational point. It is certainly easy to write down conics without rational points. For a change, let us use homogeneous co-ordinates. There is no rational point on

$$X^2 + Y^2 + Z^2 = 0,$$

since clearly there are no real points. Again, there are no rational points on

$$X^2 + Y^2 - 3Z^2 = 0. \tag{$*$}$$

For suppose (x, y, z) were such a rational point. By homogeneity, we may suppose that x, y, z are integers without common divisor. Now $(*)$ implies $x^2 + y^2 \equiv 0$ (3) and so $x \equiv y \equiv 0$ (3). Then $(*)$ gives $z \equiv 0$ (3), so x, y, z have the common factor 3: a contradiction.

For our purposes, it is convenient, and ultimately indispensable, to express the last argument in a different way. We shall introduce the fields $\mathbf{Q}_p$ of p-adic numbers, where p is a prime (here $p = 3$); and what we have just done can be expressed as proving that there are no points on $(*)$ defined over $\mathbf{Q}_3$.

2

p-adic numbers

Most of the familiar properties of the ordinary absolute value on the real or complex fields are consequences of the following three:

(i) $|r| \geq 0$, with equality precisely for $r = 0$.

(ii) $|rs| = |r||s|$.

(iii) $|r + s| \leq |r| + |s|$.

A real-valued function $|.|$ on a field k is said to be a *valuation* if it satisfies (i), (ii) (iii). Since $(-1)^2 = 1$, properties (i)-(iii) imply that $|-1| = 1$, $|-r| = |r|$ (all r).

The rational field $\mathbf{Q}$ has other valuations than the absolute value. Let p be a fixed prime. Any rational $r \neq 0$ can be put in the shape

$$r = p^\rho u/v, \ \rho \in \mathbf{Z}, \ u, \ v \in \mathbf{Z}, \ p \nmid u, \ p \nmid v.$$

We define

$$|r|_p = p^{-\rho}$$

and

$$|0|_p = 0.$$

This definition clearly satisfies (i), (ii) above. Let

$$s = p^\sigma m/n \qquad m, \ n \in \mathbf{Z}, \ p \nmid m, \ p \nmid n,$$

so

$$|s|_p = p^{-\sigma},$$

where without loss of generality

$$\sigma \geq \rho, \text{ i.e. } |s|_p \leq |r|_p.$$

Then

$$r + s = p^\rho(un + p^{\sigma-\rho}mv)/vn.$$

Here $p \nmid vn$. The numerator $un + p^{\sigma-\rho}mv$ is an integer, but, at least for for $\rho = \sigma$, it may be divisible by p. Hence

$$|r + s|_p \leq p^{-\rho},$$

that is

(iii*) $|r + s|_p \leq \max\{|r|_p, |s|_p\}$.

Clearly (iii*) implies (iii), so $|\ |_p$ is a valuation. We call it the *p-adic valuation*. The inequality (iii*) is called the *ultrametric inequality*, since (iii), the *triangle inequality*, expresses the fact that $|r - s|$ is a metric. A valuation which satisfies the ultrametric inequality is said to be *non-archimedean*.

We can transfer familiar terminology from the ordinary absolute value to the p-adic case. For example, we say that a sequence $\{a_n\}$, $n = 1, 2, \ldots$ is a *fundamental sequence* if for any $\varepsilon > 0$ there is an $n_0(\varepsilon)$ such that

$$|a_m - a_n|_p < \varepsilon \qquad \text{whenever} \qquad m, n \geq n_0(\varepsilon).$$

The sequence $\{a_n\}$ *converges* to b if

$$|a_n - b|_p < \varepsilon \qquad (\text{all } n \geq n_0(\varepsilon)).$$

For example let

$$p = 5$$

and consider the sequence

$$\{a_n\}: \quad 3, \quad 33, \quad 333, \quad 3333, \quad \ldots .$$

Then

$$a_m \equiv a_n \qquad \text{mod } 5^n \qquad (m \geq n)$$

i.e.

$$|a_m - a_n|_5 \leq 5^{-n} \qquad (m \geq n).$$

Hence $\{a_n\}$ is a fundamental sequence. Indeed it is a convergent sequence, since

$$3a_n = 99\ldots 99 \equiv -1(5^n),$$

i.e.

$$|3a_n + 1|_5 \leq 5^{-n}$$

and so

$$a_n \rightarrow -1/3$$

5-adically.

As the above example shows, the main difficulties with the p-adic valuation are psychological: something is p-adically small if it is divisible by a high power of p. Not every p-adic fundamental sequence is convergent. Let us take $p = 5$ again. Then we construct a sequence of $a_n \in \mathbf{Z}$ such that

$$a_n^2 + 1 \equiv 0 \ (5^n)$$

and

$$a_{n+1} \equiv a_n \ (5^n).$$

We start with $a_1 = 2$. Suppose that we already have a_n for some n and put $a_{n+1} = a_n + b5^n$, where $b \in \mathbf{Z}$ is to be determined. We require

$$(a_n + b5^n)^2 + 1 \equiv 0 \ (5^{n+1}),$$

that is

$$2a_n b + c \equiv 0 \ (5), \qquad\qquad (*)$$

where we already have

$$c = (a_n^2 + 1)/5^n \in \mathbf{Z}.$$

Clearly $5 \nmid a_n$ and so we can solve the congruence $(*)$ for the unknown b.

The sequence $\{a_n\}$ just constructed is a 5-adic fundamental sequence since

$$|a_m - a_n|_5 \leq 5^{-n} \qquad (m \geq n).$$

Suppose, if possible, that a_n tends 5-adically to some $e \in \mathbf{Q}$. Then

$$a_n^2 + 1 \to e^2 + 1.$$

On the other hand, by our construction,

$$a_n^2 + 1 \to 0.$$

Hence $e^2 + 1 = 0$; a contradiction.

Just as the real numbers are constructed by completing the rationals with respect to the ordinary absolute value, so the rationals can be completed with respect to $| \ |_p$ to give the field $\mathbf{Q}_p$ of *p-adic numbers*. In fact the process can be simplified because $| \ |_p$ is non-archimedean. For the reader who is unfamiliar with this way of constructing the reals, we sketch a construction of $\mathbf{Q}_p$ at the end of this section.

We say that a field K is *complete* with respect to a valuation $|.|$ if every fundamental sequence is convergent. A field K with valuation $||.||$ is said to be the *completion* of the field k with valuation $|.|$ if there is an injection

$$\lambda : k \to K$$

which preserves the valuation:

$$\|\lambda a\| = |a| \qquad (a \in k)$$

and such that

(i) K is complete with respect to $\|.\|$

(ii) K is the closure of λk with respect to the topology induced by $\|.\|$
      (K is not "too large").

The completion always exists and is unique (up to a unique isomorphism). We henceforth identify k with λk and $|.|$ with $\|.\|$, so regard k as a subfield of K.

We now discuss the structure of the p-adic field $\mathbf{Q}_p$ with its valuation $|\ |_p$.

We note that

$$|a + b|_p = |a|_p \qquad \text{if} \qquad |b|_p < |a|_p.$$

For by (iii*) $|a + b|_p \leq |a|_p$ and, since $a = (a + b) + (-b)$, we have a contradiction if $|a + b|_p < |a|_p$. It follows that the set of values taken by $|\ |_p$ on $\mathbf{Q}_p$ is precisely the same as the set for $\mathbf{Q}$. Indeed if $\alpha \in \mathbf{Q}_p$, $\alpha \neq 0$ then by (ii) of the definition of the completion, there is an $a \in \mathbf{Q}$ with $|a - \alpha|_p < |\alpha|_p$, so $|\alpha|_p = |a|_p$.

The set of $\alpha \in \mathbf{Q}_p$ with $|\alpha| \leq 1$ is called the set of *p-adic integers* $\mathbf{Z}_p$. Because $|\ |_p$ is non-archimedean, $\mathbf{Z}_p$ is a ring:

$$|\alpha|_p, \ |\beta|_p \leq 1 \Rightarrow |\alpha\beta|_p \leq 1, \ |\alpha + \beta|_p \leq 1.$$

A rational number b is in $\mathbf{Z}_p$ precisely when it has the form $b = u/v$, where $u, v \in \mathbf{Z}$, $p \nmid v$.

The numbers $\varepsilon \in \mathbf{Q}_p$ with $|\varepsilon| = 1$ are the *p-adic units*. From what was said about the values taken by $|.|_p$ on $\mathbf{Q}_p$, every $\beta \neq 0$ in $\mathbf{Q}_p$ is of the shape $\beta = p^n \varepsilon$, where $n \in \mathbf{Z}$ and ε is a unit. The units are just the elements ε of $\mathbf{Q}_p$ such that $\varepsilon \in \mathbf{Z}_p$, $\varepsilon^{-1} \in \mathbf{Z}_p$.

As we have already noted, elementary analysis continues to hold in $\mathbf{Q}_p$, but can be simpler; as the following lemma shows.

Lemma 1. *In $\mathbf{Q}_p$ the series $\sum_0^\infty \beta_n$ converges if and only if $\beta_n \to 0$.*

Proof. By saying that the sum converges, we mean, of course, that the partial sums $\sum_0^N$ tend to a limit.

That convergence implies $\beta_n \to 0$ is true even in real analysis. To prove the opposite implication, we note that

$$\left| \sum_0^N - \sum_0^M \right|_p = \left| \sum_{M+1}^N \beta_n \right|_p$$

$$\leq \max_{M<n\leq N} |\beta_n|_p$$

by an obvious extension of the ultrametric inequality (iii*) to several summands. Hence $\left\{ \sum_0^N \beta_n \right\}$ is a fundamental sequence, so tending to a limit by the completeness of $\mathbf{Q}_p$.

We are now in a position to give an explicit description of $\mathbf{Z}_p$. We write

$$\mathcal{A} = \{0, 1, \ldots, p-1\}.$$

Lemma 2. *The elements of* $\mathbf{Z}_p$ *are precisely the sums*

$$\alpha = \sum_0^\infty a_n p^n,$$

where

$$a_n \in \mathcal{A} \qquad \text{(all } n\text{)}.$$

Proof. By the preceeding lemma, the infinite sum converges, and its value is clearly in $\mathbf{Z}_p$.

Now let $\alpha \in \mathbf{Z}_p$ be given. There is a $b \in \mathbf{Q}$ such that $|b - \alpha|_p < 1$, and it is easy to prove that there is precisely one $a_0 \in \mathcal{A}$ such that $|a_0 - b|_p < 1$. Then

$$\alpha = a_0 + p\alpha_1$$

where $|\alpha_1| \leq 1$, i.e. $\alpha_1 \in \mathbf{Z}_p$. Proceeding inductively, we get

$$\alpha = a_0 + a_1 p + \ldots + a_N p^N + \alpha_N p^{N+1}$$

with $\alpha_N \in \mathbf{Z}_p$.

For the final result we must distinguish between $p = 2$ and $p \neq 2$.

Lemma 3 ($p \neq 2$). *Let* $\alpha \in \mathbf{Q}_p$ *be a unit. A necessary and sufficient condition that* $\alpha = \beta^2$ *for some* $\beta \in \mathbf{Q}_p$ *in that there is some* $\gamma \in \mathbf{Q}_p$ *with*

$$|\alpha - \gamma^2|_p < 1.$$

Proof. Necessity is obvious. We have already in effect given a proof in the special case $p = 5$, $\alpha = -1$. That in the general case is similar: one constructs inductively $\beta_1 = \gamma$, β_2, β_3, ... such that

$$|\beta_n^2 - \alpha| \leq p^{-n}$$

$$|\beta_{n+1} - \beta_n| \leq p^{-n}$$

If we already have β_n, we take $\beta_{n+1} = \beta_n + \delta$, so

$$\beta_{n+1}^2 = \beta_n^2 + 2\beta_n\delta + \delta^2$$

and it is enough to take

$$\delta = (\alpha - \beta_n^2)/2\beta_n.$$

This lemma ceases to hold for $p = 2$ (consider $\alpha = 5$, $\beta = 1$). We have

Lemma 4 $(p = 2)$. *Let* $\alpha \in \mathbf{Q}_2$ *be a unit. A necessary and sufficient condition that* $\alpha = \beta^2$ *for some* $\beta \in \mathbf{Q}_2$ *is that* $|\alpha - 1| \leq 2^{-3}$.

Proof. Here again, the necessity is obvious. For sufficiency we construct a sequence $\beta_1 = 1$, β_2, β_3, ... as in the previous proof. The details are left to the reader.

We conclude this section by the promised sketch of the construction of $\mathbf{Q}_p$.

Denote by $\mathfrak{F}$ the set of fundamental sequences $\{a_n\}$ for $|\ |_p$, where $a_n \in \mathbf{Q}$. Then $\mathfrak{F}$ is a ring under componentwise addition and multiplication.

$$\{a_n\} + \{b_n\} = \{a_n + b_n\} : \{a_n\}\{b_n\} = \{a_nb_n\}.$$

A sequence $\{a_n\}$ is a null sequence if $a_n \to 0$ (p-adically). The set $\mathfrak{N}$ of null-sequences is clearly an ideal in $\mathfrak{F}$.

Let $\{a_n\} \in \mathfrak{F}$ but $\{a_n\} \notin \mathfrak{N}$. Then it is easy to see that there is at least one N such that $|a_N - a_n| < |a_N|_p$ for all $n > N$. Then $|a_n|_p = |a_N|_p$ for all $n \geq N$. We write $|\{a_n\}|_p = |a_N|_p$. If $a_n \neq 0$ for all n, it is now easy to deduce that $\{a_n^{-1}\} \in \mathfrak{F}$.

We show that $\mathfrak{N}$ is a maximal ideal in $\mathfrak{F}$. For, if not, let $\mathfrak{M}$ be a strictly bigger ideal than $\mathfrak{N}$. It must contain an $\{a_n\} \notin \mathfrak{N}$. Then only finitely many of the a_n can be 0, and replacing them by (say) 1 merely adds an element of $\mathfrak{N}$. Hence we can suppose that $a_n \neq 0$ for all n. Then $\{a_n^{-1}\} \in \mathfrak{F}$, and so $\{a_n^{-1}\}\{a_n\} \in \mathfrak{M}$. Hence we should have $\mathfrak{M} = \mathfrak{F}$, a contradiction. We conclude that $\mathfrak{N}$ is maximal, and thus $\mathfrak{F}/\mathfrak{N}$ is a field.

The field $\mathbf{Q}$ is mapped into $\mathfrak{F}/\mathfrak{N}$ by

$$r \to \{r\} \in \mathfrak{F}.$$

The function $|\{a_n\}|$ on $\mathfrak{F}$ induces a function on $\mathfrak{F}/\mathfrak{N}$ which is easily seen to be a valuation and to coincide with $|\ |_p$ on the image of $\mathbf{Q}$.

Finally, it is not difficult to check that $\mathfrak{F}/\mathfrak{N}$ is itself complete by a diagonal argument on a sequence of elements of $\mathfrak{F}$.

§2. Exercises

1. For each of the sets of p, m, r given, either find an $x \in \mathbb{Z}$ such that
$$|r - x|_p \leq p^{-m},$$
or show that no such x exists.

(i) $p = 257$, $r = 1/2$, $m = 1$;
(ii) $p = 3$, $r = 7/8$, $m = 2$;
(iii) $p = 3$, $r = 7/8$, $m = 7$;
(iv) $p = 3$, $r = 5/6$, $m = 9$;
(v) $p = 5$, $r = 1/4$, $m = 4$.

2. Construct further examples along the lines of Exercise 1 until the whole business seems trivial.

3. For given p, m, r either find an $x \in \mathbb{Z}$ such that
$$|r - x^2|_p \leq p^{-m}$$
or show that no such x exists.

(i) $p = 5$, $r = -1$, $m = 4$;
(ii) $p = 5$, $r = 10$, $m = 3$;
(iii) $p = 13$, $r = -4$, $m = 3$;
(iv) $p = 2$, $r = -7$, $m = 6$;
(v) $p = 7$, $r = -14$, $m = 4$;
(vi) $p = 7$, $r = 6$, $m = 3$;
(vii) $p = 7$, $r = 1/2$, $m = 3$.

4. As Exercise 2.

5. Let $p > 0$ be prime, $p \equiv 2\ (3)$. For any integer a, $p \nmid a$, show that there is an $x \in \mathbb{Z}_p$ with $x^3 = a$.

3

The local-global principle for conics

We have seen that the theory of curves of genus 0 over $\mathbf{Q}$ turns on deciding whether a given conic has a rational point.

We use homogeneous co-ordinates. A conic C defined over $\mathbf{Q}$ is given by an equation

$$F(\mathbf{X}) = \sum f_{ij} X_i X_j = 0$$

where $\mathbf{X} = (X_1, X_2, X_3)$,

$$f_{ij} = f_{ji} \in \mathbf{Q}$$

and the quadratic form F (recall a *form* is a homogeneous polynomial) is nonsingular, i.e.

$$\det(f_{ij}) \neq 0.$$

In our initial discussion we noted that, apart from reality considerations, we could disprove the existence of rational points by congruence considerations. These we now replace by reference to p-adic numbers.

A criterion for the existence of a rational point on a conic was given by Legendre. It was left to Hasse to give it the following succinct formulation.

Theorem 1. *A necessary and sufficient condition for the existence of a rational point on a conic C defined over $\mathbf{Q}$ is that there is a point defined over the real field $\mathbf{R}$ and over $\mathbf{Q}_p$ for every prime p.*

Necessity is trivial. We shall prove sufficiency, but it will require some time and preparation. First we introduce some conventional terminology.

The real field $\mathbf{R}$ is somewhat analogous to the $\mathbf{Q}_p$ and is conventionally denoted by $\mathbf{Q}_\infty$. When we write $\mathbf{Q}_p$ we will not include $p = \infty$ unless we explicitly say so. The fields $\mathbf{Q}_p$ (including $p = \infty$) are called the localizations of $\mathbf{Q}$. In contrast, $\mathbf{Q}$ is called the global field. We say that something is true "everywhere locally" if it is true for all $\mathbf{Q}_p$ (including ∞). In this lingo the theorem becomes "A necessary and sufficient condition for the existence of a global point on a conic is that there should be a point everywhere locally".

The local-global theorem for conics implies a local-global theorem for curves of genus 0 but some care must be taken in the formulation ["point" must be interpreted as "place"]. We do not pursue this further.

In the rest of this section we transform the theorem into a shape better suited for attack[1].

A transformation

$$T: \quad X_i = \sum_i t_{ij} Y_j$$

with

$$t_{ij} \in \mathbf{Q}, \qquad \det(t_{ij}) \neq 0$$

takes the quadratic form $F(\mathbf{X})$ into a quadratic form $G(\mathbf{Y})$, say. Then T takes points defined over $\mathbf{Q}$ on $F(\mathbf{X}) = 0$ into points defined over $\mathbf{Q}$ on $G(\mathbf{Y}) = 0$ and, similarly, the inverse T^{-1} takes points on $G(\mathbf{Y}) = 0$ to points on $F(\mathbf{X}) = 0$. Likewise for points defined over $\mathbf{Q}_p$ for each p (including ∞). Hence the theorem holds for $F(\mathbf{X}) = 0$ if and only if it holds for $G(\mathbf{Y}) = 0$.

By suitable choice of transformation T we thus need consider only "diagonal" forms

$$F(\mathbf{X}) = f_1 X_1^2 + f_2 X_2^2 + f_3 X_3^2.$$

By substitutions $X_j \rightarrow t_j X_j$ $(t_j \in \mathbf{Q})$ we may suppose without loss of generality that the

$$f_j \in \mathbf{Z}$$

are square free.

If f_1, f_2, f_3 have a prime factor p in common, we replace $F(\mathbf{X})$ by $p^{-1}F(\mathbf{X})$. If two of the f_j, say f_1, f_2 have a prime p in common but $p \nmid f_3$, we replace X_3 by pX_3 and then divide F by p. Both of these

---

[1] The details of the proof of Theorem 1 will not be required for the treatment of elliptic curves. The reader who is interested only in the latter should omit the rest of this § and also omit §§4,5.

transformations reduce the absolute value of the integer $f_1f_2f_3$. After a finite number of steps we are reduced to the case when $f_1f_2f_3$ is square free. We have thus proved the

Metalemma 1. *To prove the Theorem, it is enough to prove it for conics*

$$F(\mathbf{X}) = f_1X_1^2 + f_2X_2^2 + f_3X_3^2 = 0,$$

where $f_j \in \mathbf{Z}$ and $f_1f_2f_3$ is square free.

The next stage is to draw conclusions from the hypothesis that a conic as described in the Metalemma has points everwhere locally. There is a point defined over $\mathbf{Q}_p$ when there is a vector $\mathbf{a} = (a_1, a_2, a_3) \neq (0,0,0)$ with $a_j \in \mathbf{Q}_p$ such that $F(\mathbf{a}) = 0$. By multiplying the a_j by an element of $\mathbf{Q}_p$ we may suppose without loss of generality that

$$\max |a_j|_p = 1. \qquad (*)$$

For our later purposes we have to consider several cases.

First case. $p \neq 2, p \mid f_1f_2f_3$. Without loss of generality $p \mid f_1$, so $p \nmid f_2$, $p \nmid f_3$. Then $|f_1a_1^2|_p < 1$. Suppose, if possible that $|a_2|_p < 1$. Then

$$|f_3a_3^2|_p = |f_1a_1^2 + f_2a_2^2|_p < 1$$

and $|a_3|_p < 1$. Now

$$|f_1a_1^2|_p = |f_2a_2^2 + f_3a_3^2|_p \leq p^{-2}$$

and so $|a_1|_p < 1$ since f_1 is square free. This contradicts the normalization $(*)$, and so $|a_2|_p = |a_3|_p = 1$. But now

$$|f_2a_2^2 + f_3a_3^2|_p < 1.$$

On dividing by the unit a_2, we deduce that there is some $r_p \in \mathbf{Z}$ such that

$$f_2 + r_p^2 f_3 \equiv 0 \ (p).$$

Second case. $p = 2, 2 \nmid f_1f_2f_3$. It is easy to see that precisely two of the a_j are units, say a_2 and a_3. Now $a^2 \equiv 1$ or 0 (4) for $a \in \mathbf{Z}$; and so

$$f_2 + f_3 \equiv 0 \ (4).$$

Third case. $p = 2, 2 \mid f_1f_2f_3$, say $2 \mid f_1$. Now $|a_2|_2 = |a_3|_3 = 1$. Now $a^2 \equiv 1$ (8) for $a \in \mathbf{Z}, 2 \nmid a$; and so

$$f_2 + f_3 \equiv 0 \ (8)$$

or

$$f_1 + f_2 + f_3 \equiv 0 \ (8)$$

according as $|a_1|_2 < 1$ or $|a_1|_2 = 1$.

In the next two sections, we show that the conditions just derived are sufficient to ensure the existence of a global point on $F(\mathbf{X}) = 0$.

§3. Exercises

1. (i) Let $p > 2$ be prime and let $b, c \in \mathbf{Z}$, $p \nmid b$. Show that $bx^2 + c$ takes precisely $\frac{1}{2}(p+1)$ distinct values p for $x \in \mathbf{Z}$. (ii) Suppose that, further, $a \in \mathbf{Z}$, $p \nmid a$. Show that there are $x, y \in \mathbf{Z}$ such that $bx^2 + c \equiv ay^2$ (p).

2. Let $a, b, c \in \mathbf{Z}_p$, $|a|_p = |b|_p = |c|_p = 1$ where p is prime, $p > 2$. Show that there are $x, y \in \mathbf{Z}_p$ such that $bx^2 + c = ay^2$.

3. Let $p > 2$ be prime, $a_{ij} \in \mathbf{Z}$ $(1 \le i, j \le 3)$, $a_{ji} = a_{ij}$ and let $d = \det(a_{ij})$. Suppose that $p \nmid d$. Show that there are $x_1, x_2, x_3 \in \mathbf{Z}$, not all divisible by p, such that $\sum_{i,j} a_{ij} x_i x_j \equiv 0$ (p).

4. Let $a, b, c \in \mathbf{Z}$, $2 \nmid abc$. Show that a necessary and sufficient condition that the only solution in $\mathbf{Q}_2$ of $ax^2 + by^2 + cz^2 = 0$ is the trivial one is that $a \equiv b \equiv c$ (4).

5. For each of the following sets of a, b, c find the set of primes p (including ∞) for which the only solution of $ax^2 + by^2 + cz^2 = 0$ in $\mathbf{Q}_p$ is the trivial one:

(i) $(a, b, c) = (1, 1, -2)$
(ii) $(a, b, c) = (1, 1, -3)$
(iii) $(a, b, c) = (1, 1, 1)$
(iv) $(a, b, c) = (14, -15, 33)$

6. Do you observe anything about the parity of the number N of primes (including ∞) for which there is insolubility? If not, construct similar exercises and solve them until the penny drops.

7.(i) Prove your observation in (6) in the special case $a = 1$, $b = -r$, $c = -s$, where r, s are distinct primes > 2.
[Hint. Quadratic reciprocity]
(ii)   [Difficult]. Prove your observation for all $a, b, c \in \mathbf{Z}$.

4

Geometry of numbers

At this stage we require a tool from the Geometry of Numbers, which we shall develop from scratch.

A generalization of the pigeon-hole principle (Schubfachprinzip) says that if we have N things to file in H holes and $N > mH$ for an integer m, then at least one of the holes will contain $\geq (m+1)$ things. We start with a continuous analogue.

Let $\mathbf{R}^n$ denote the vector space of real n-tuples $\mathbf{r} = (r_1, \ldots, r_n)$. It contains the group $\mathbf{Z}^n$ of $\mathbf{r}$ for which $r_j \in \mathbf{Z}$ (all j). By the volume $V(\mathcal{S})$ of a set $\mathcal{S} \subset \mathbf{R}^n$ we shall mean its Lebesgue measure, but in the applications we will be concerned only with very simple-minded $\mathcal{S}$.

Lemma 1. *Let $m > 0$ be an integer and let $\mathcal{S} \subset \mathbf{R}^n$ with*

$$V(\mathcal{S}) > m.$$

Then there are $m + 1$ distinct points $\mathbf{s}_0, \ldots, \mathbf{s}_m$ of S such that

$$\mathbf{s}_i - \mathbf{s}_j \in \mathbf{Z}^n \qquad (0 \leq i,\ j \leq m).$$

Proof. Let $\mathcal{W} \subset \mathbf{R}^n$ be the "unit cube" of points $\mathbf{w}$ with

$$0 \leq w_j < 1 \qquad (1 \leq j \leq n).$$

Then every $\mathbf{x} \in \mathbf{R}^n$ is uniquely of the shape

$$\mathbf{x} = \mathbf{w} + \mathbf{z},$$

where $\mathbf{z} \in \mathbf{Z}^n$. Let $\psi(\mathbf{x})$ be the characteristic function of S $(= 1$ if $\mathbf{x} \in \mathcal{S}$,

$= 0$ otherwise). Then

$$m < V(\mathcal{S}) = \int_{\mathbf{R}^n} \psi(\mathbf{x})d\mathbf{x}$$

$$= \int_{\mathcal{W}} \left(\sum_{\mathbf{z} \in \mathbf{Z}^n} \psi(\mathbf{w} + \mathbf{z}) \right) d\mathbf{w}.$$

Since $V(\mathcal{W}) = 1$, there must be some $\mathbf{w}_0 \in \mathcal{W}$ such that

$$\sum_{\mathbf{z} \in \mathbf{Z}^n} \psi(\mathbf{w}_0 + \mathbf{z}) > m,$$

$$\text{so} \quad \geq m + 1.$$

We may now take for the s_j the $\mathbf{w}_0 + \mathbf{z}$ for which $\psi(\mathbf{w}_0 + \mathbf{z}) > 0$.

The set $\mathcal{S}$ is said to be *symmetric* (about the origin) if $-\mathbf{x} \in \mathcal{S}$ whenever $\mathbf{x} \in \mathcal{S}$. It is *convex* if whenever $\mathbf{x}$, $\mathbf{y} \in \mathcal{S}$, then the whole line-segment

$$\lambda \mathbf{x} + (1 - \lambda)\mathbf{y} \in \mathcal{S} \qquad (0 \leq \lambda \leq 1)$$

joining them is in $\mathcal{S}$. In particular, the mid-point $\frac{1}{2}(\mathbf{x} + \mathbf{y})$ is in $\mathcal{S}$.

Theorem 1. *Let Λ be a subgroup of $\mathbf{Z}^n$ of index m. Let $\mathcal{C} \subset \mathbf{R}^n$ be a symmetric convex set of volume*

$$V(\mathcal{C}) > 2^n m.$$

Then $\mathcal{C}$ and Λ have a common point other than $\mathbf{0} = (0, \ldots, 0)$.

Proof. Let $\mathcal{S} = \frac{1}{2}\mathcal{C}$ be the set of points $\frac{1}{2}\mathbf{c}$, $\mathbf{c} \in \mathcal{C}$. Then

$$V(\frac{1}{2}\mathcal{C}) = 2^{-n}V(\mathcal{C}) > m.$$

By Lemma 1, there are $m + 1$ distinct points $\mathbf{c}_0, \ldots, \mathbf{c}_m \in \mathcal{C}$ such that

$$\frac{1}{2}\mathbf{c}_i - \frac{1}{2}\mathbf{c}_j \in \mathbf{Z}^n . \qquad (0 \leq i, j \leq m).$$

There are $m + 1$ points

$$\frac{1}{2}\mathbf{c}_i - \frac{1}{2}\mathbf{c}_0 \qquad (0 \leq i \leq m)$$

and m cosets of $\mathbf{Z}^n$ modulo Λ. By the pigeon hole principle, two must be in the same coset, that is there are i, j with $i \neq j$ such that

$$\frac{1}{2}\mathbf{c}_i - \frac{1}{2}\mathbf{c}_j \in \Lambda.$$

Now $-\mathbf{c}_j \in \mathcal{C}$ by symmetry; and so

$$\frac{1}{2}\mathbf{c}_i - \frac{1}{2}\mathbf{c}_j = \frac{1}{2}\mathbf{c}_i + \frac{1}{2}(-\mathbf{c}_j) \in \mathcal{C}$$

by convexity.

Note. Lemma 1 and Theorem 1 with $m = 1$ are due to Blichfeldt and Minkowski respectively. The generalizations to $m > 1$ are by van der Corput.

As a foretaste of the flavour of the application in the next section, we give

Lemma 2. *Let N be a positive integer. Suppose that there is an $l \in \mathbf{Z}$ such that*

$$l^2 \equiv -1 \ (N).$$

Then $N = u^2 + v^2$ for some $u, v \in \mathbf{Z}$.

Proof. We take $n = 2$ and denote the co-ordinates by x, y. For $\mathcal{C}$ we take the open disc

$$x^2 + y^2 < 2m$$

of volume (= area)

$$V(\mathcal{C}) = 2\pi m > 2^2 m.$$

The subgroup Λ of $\mathbf{Z}^2$ is given by

$$x, y \in \mathbf{Z}, \qquad y \equiv lx \ (m).$$

It is clearly of index m. Hence by the Theorem there is

$$(0,0) \neq (u,v) \in \Lambda \cap \mathcal{C}.$$

Then

$$0 < u^2 + v^2 < 2m$$

and

$$u^2 + v^2 \equiv u^2(1 + l^2) \equiv 0 \ (m).$$

Hence $u^2 + v^2 = m$, as required.

We note, in passing, that the condition of the lemma is certainly satisfied for primes p with $p \equiv 1 \ (4)$.

§4. Exercises

1. Let $m \in \mathbf{Z}$, $m > 1$ and suppose that there is some $f \in \mathbf{Z}$ such that $f^2 + f + 1 \equiv 0 \ (m)$. Show that $m = u^2 + uv + v^2$ for some $u, v \in \mathbf{Z}$.

2. Find a prime $p > 0$ for which there is an $f \in \mathbf{Z}$ such that

$$1 + 5f^2 \equiv 0 \ (p)$$

but p is not of the shape $u^2 + 5v^2$ ($u, v \in \mathbf{Z}$).

5

Local-global principle. Conclusion of proof

We now complete the proof of the local-global principle for conics using the theorem of the last section. We recall that we had reduced the proof to that for

$$f_1 X_1^2 + f_2 X_2^2 + f_3 X_3^2 = 0$$

where f_1, f_2, $f_3 \in \mathbf{Z}$ and $f_1 f_2 f_3$ is square free. We assume that there are points everywhere locally and we showed that this implied certain congruences to primes p dividing $2 f_1 f_2 f_3$.

We first define a subgroup Λ of $\mathbf{Z}^3$ by imposing congruence conditions on the components of $\mathbf{x} = (x_1, x_2, x_3)$.

First case. $p \neq 2$, $p \nmid f_1 f_2 f_3$, say $p \mid f_1$. We saw (end of §3) that then there is an $r_p \in \mathbf{Z}$ and that

$$f_2 + r_p^2 f_3 \equiv 0 \quad (p).$$

We impose the condition

$$x_3 \equiv r_p x_2 \quad (p).$$

Then

$$
\begin{aligned}
F(\mathbf{x}) &= f_1 x_1^2 + f_2 x_2^2 + f_3 x_3^2 \\
&\equiv (f_2 + r_p^2 f_3) x_2^2 \\
&\equiv 0 \quad (p).
\end{aligned}
$$

Second case. $p = 2$, $2 \nmid f_1 f_2 f_3$. Then without loss of generality

$$f_2 + f_3 \equiv 0 \quad (4).$$

We impose the conditions

$$\left.\begin{aligned} x_1 &\equiv 0 \quad (2) \\ x_2 &\equiv x_3 \quad (2) \end{aligned}\right\},$$

which imply

$$F(\mathbf{x}) \equiv 0 \quad (4).$$

Third case. $p = 2$, $2 \mid f_1 f_2 f_3$, say $2 \mid f_1$. Then

$$s^2 f_1 + f_2 + f_3 \equiv 0 \quad (8),$$

where $s = 0$ or 1. We impose the conditions

$$\left.\begin{aligned} x_2 &\equiv x_3 \quad (4) \\ x_1 &\equiv s x_3 \quad (2) \end{aligned}\right\},$$

which imply

$$F(\mathbf{x}) \equiv 0 \quad (8).$$

To sum up. The group Λ is of index m (say) $= 4|f_1 f_2 f_3|$ in $\mathbf{Z}^3$, where throughout this section $|\ |$ is the absolute value. Further,

$$F(\mathbf{x}) \equiv 0 \qquad (4\,|f_1 f_2 f_3|)$$

for $\mathbf{x} \in \Lambda$.

We apply the theorem of the previous section to Λ and the convex symmetric set

$$\mathcal{C} : |f_1| x_1^2 + |f_2| x_2^2 + |f_3| x_3^2 < 4|f_1 f_2 f_3|.$$

School geometry shows that

$$\begin{aligned} V(\mathcal{C}) &= (\pi/3).2^3.|4 f_1 f_2 f_3| \\ &> 2^3 |4 f_1 f_2 f_3| \\ &= m. \end{aligned}$$

Hence there is an $\mathbf{c} \neq \mathbf{0}$ in $\Lambda \cap \mathcal{C}$. For this $\mathbf{x}$ we have

$$F(\mathbf{x}) \equiv 0 \quad (4|f_1 f_2 f_3|)$$

and

$$\begin{aligned} |F(\mathbf{x})| &\leq |f_1| x_1^2 + |f_2| x_2^2 + |f_3| x_3^2 \\ &< 4|f_1 f_2 f_3|; \end{aligned}$$

so

$$F(\mathbf{x}) = 0,$$

as required.

We conclude with some remarks.

Remark 1. We have not merely shown that there is a solution of $F(\mathbf{x}) = 0$, but we have found that there is one in a certain ellipsoid. This facilitates the search in explicitly given cases.

Remark 2. We have made no use of the condition of solubility in $\mathbf{Q}_p$ for $p \nmid 2f_1 f_2 f_3$. In fact this condition tells us nothing [cf. §3, Exercises 2, 3]. It is left to the reader to check that for any f_1, f_2, f_3 and p with $p \nmid 2f_1 f_2 f_3$ there is always a point defined over $\mathbf{Q}_p$ on

$$f_1 X_1^2 + f_2 X_2^2 + f_3 X_3^2 = 0.$$

Remark 3. We have also nowhere used that there is local solubility for $\mathbf{Q}_\infty = \mathbf{R}$.

Hence solubility at $\mathbf{Q}_\infty$ is implied by solubility at all the $\mathbf{Q}_p$ $(p \neq \infty)$. This phenomenon is connected with quadratic reciprocity. In fact for any conic over $\mathbf{Q}$, the number of p (including ∞) for which there is not a point over $\mathbf{Q}_p$ is always even [cf. §3, Exercises 6,7]. See a book on quadratic forms (such as the author's).

§5. Exercises

1. Let

$$F(X, Y, Z) = 5X^2 + 3Y^2 + 8Z^2 + 6(YZ + ZX + XY).$$

Find rational integers x, y, z not all divisible by 13, such that

$$F(x, y, z) \equiv 0 \ (\mathrm{mod} \ 13^2).$$

[*Hint.* cf. Hensel's Lemma 2 of §10.]

2. Let

$$F(X, Y, Z) = 7X^2 + 3Y^2 - 2Z^2 + 4YZ + 6ZX + 2XY.$$

Find rational integers x, y, z not all divisible by 17 such that

$$F(x, y, z) \equiv 0 \ (\mathrm{mod} \ 17^3).$$

6

Cubic curves

In this section we consider curves given by

$$\mathcal{C} : F(\mathbf{X}) = F(X_1, X_2, X_3) = 0,$$

where F is a homogeneous cubic form. The case of interest is when the ground field is the rationals $\mathbf{Q}$, but our initial remarks apply to any ground field.

A point $\mathbf{x}$ on $\mathcal{C}$ is said to be *singular* when

$$\frac{\partial F}{\partial X_j}(\mathbf{x}) = 0 \qquad (j = 1, 2, 3).$$

If we choose co-ordinates so that $\mathbf{x} = (0, 0, 1)$, this is equivalent to F not containing terms in X_3^3, $X_1 X_3^2$, $X_2 X_3^2$.

A singular point counts with multiplicity at least 2 as an intersection with a line. More precisely, if $\mathbf{a}$, $\mathbf{b}$ are two points on the line, the general point on it is

$$\lambda\mathbf{a} + \mu\mathbf{b},$$

where the numbers λ, μ are not both 0. The intersections with $\mathcal{C}$ are given by

$$F(\lambda\mathbf{a} + \mu\mathbf{b}) = 0, \tag{$*$}$$

a homogeneous cubic in λ, μ. What is claimed is that if one of the intersections is a singular point of $\mathcal{C}$ then the corresponding ratio $\lambda : \mu$ occurs as a multiple root of ($*$). An easy way to check this is to take $\mathbf{b} = \mathbf{x}$.

Suppose that C has two distinct singular points $\mathbf{x}$, $\mathbf{y}$. The line joining them cuts C at both $\mathbf{x}$, $\mathbf{y}$ with multiplicity ≥ 2. This can happen only if $F(\lambda\mathbf{x} + \mu\mathbf{y})$ vanishes identically, i.e. if C contains the whole line. If we suppose, as we shall, that C is *irreducible* (i.e. that F does not factorize), this cannot happen. An irreducible cubic curve has at most one singular point.

Now take the ground field to be $\mathbf{Q}$. If there is a singular point over the algebraic closure $\overline{\mathbf{Q}}$, there is at most one. By Galois theory[2] it must be defined over $\mathbf{Q}$. Hence, as we have already seen in §1, C is birationally equivalent over $\mathbf{Q}$ to the line.

From now on we restrict attention to *non-singular* cubic curves, i.e. those which have no singular points over $\mathbf{Q}$. Let $\mathbf{a}$, $\mathbf{b}$ be rational points on C. The line joining them meets C in a third point, in general distinct: it is also rational since it is given by a cubic equation, two of whose roots are rational. The variant in which one takes the third point of intersection of the tangent at a rational point was used already by Diophantos to find new unobvious points from obvious ones. The general process was, according to Weil, first noted by Newton. An older generation of mathematicians refer to these as the "chord and tangent processes".

In general, starting from one rational point $\mathbf{a}$ on C one obtains infinitely many by the chord and tangent processes. If this is not the case, $\mathbf{a}$ is said to be *exceptional*. For example we have

Lemma 1. *Let $a \geq 1$ be a cubic-free integer and let*
$$C: \quad X^3 + Y^3 - aZ^3 = 0.$$
The point $(1, -1, 0)$ is exceptional. For $a = 1$ the points $(0, 1, 1)$, $(1, 0, 1)$ are also exceptional. For $a = 2$ the point $(1, 1, 1)$ is exceptional. No other rational point is exceptional.

Proof. We first show that the given points are indeed exceptional. The tangent at $(1, -1, 0)$ is $X + Y = 0$, which meets C only at $(1, -1, 0)$. The other cases for $a = 1$ are similar. The tangent at $(1, 1, 1)$ for $a = 2$ is $X + Y - 2Z = 0$, which meets C again only at $(1, -1, 0)$.

Let $\mathbf{x} = (x, y, z)$ be a rational point other than those named. We may

---

[2] *For the cognoscenti. If the ground field is not perfect, the conclusion does not necessarily hold. See Note at end of §9.*

suppose that x, y, z are integers without common factor. The equation for C implies that then x, y, z are coprime in pairs.

Let $\mathbf{x_1} = (x_1, y_1, z_1)$ be the third point of intersection, where again x_1, y_1, z_1 are integers without common factor. It may be verified[3] that

$$x_1 : y_1 : z_1 = x(x^3 + 2y^3) : -y(2x^3 + y^3) : z(x^3 - y^3)$$

Let d be the greatest common divisor of the three terms on the right hand side. If a prime p divides both x and d it must also divide y, a contradiction. Hence d divides $x^3 + 2y^3$ and $2x^3 + y^3$. It thus divides $3x^3$ and $3y^3$, so $d = 1$ or 3. Hence

$$z_1 = \pm z(x^3 - y^3) \qquad \text{or} \qquad z_1 = \pm z(x^3 - y^3)/3.$$

In either case, it is readily verified that $|z_1| > |z|$ except for the $\mathbf{x}$ listed in the enunciation. By repeating the tangent process we thus get a sequence of points $\mathbf{x}$, $\mathbf{x_1}$, $\mathbf{x_2}$, ... with

$$|z| < |z_1| < |z_2| < \cdots .$$

Hence the $\mathbf{x}_j$ are distinct, and $\mathbf{x}$ is not exceptional.

§6. Exercises

1. (i) Show that the cubic curve
$$Y^2 Z = X^3 + AXZ^2 + BZ^3$$
is non-singular provided that
$$4A^3 + 27B^2 \neq 0.$$

(ii) If $4A^3 + 27B^2 = 0$, find a singularity and decide whether it is a cusp or a double point with distinct tangents.

2. (i) Let
$$F(\mathbf{x}) = a_1 X_1^3 + a_2 X_2^3 + a_3 X_3^3 + dX_1 X_2 X_3,$$
where
$$a_1 a_2 a_3 \neq 0.$$
Show that $F(\mathbf{x}) = 0$ is non-singular provided that
$$a_1 a_2 a_3 + d^3 \neq 0.$$

(ii) If $a_1 = a_2 = a_3 = 1$, $d = -3$, show that any point (x_1, x_2, x_3) with $a_1^3 = x_2^3 = x_3^3 = x_1 x_2 x_3 = 1$ is a singularity.

---

[3] This is essentially a special case of elegant formulae of Desboves for the chord and tangent processes. See Exercises and Formulary.

(iii) How does the result of (ii) square with the result proved in the text that a cubic curve has at most one singularity?

3. Let $F(x)$ be as in the previous question and suppose that $F(x) = 0$ is non-singular.

(i)   Let $F(x) = 0$. Show that the third intersection t of the tangent at x is given by

$$t_j = x_j(a_{j+1}x_{j+1}^3 - a_{j+2}x_{j+2}^3) \quad (j = 1, 2, 3),$$

where the suffixes are taken mod 3.

(ii)   Let x, y be distinct points on $F(X) = 0$. Show that the third intersection z of the line joining them is given by

$$z_j = x_j^2 y_{j+1} y_{j+2} - y_j^2 x_{j+1} x_{j+2}.$$

[Formulae of Desboves].

4. Starting with the solution $(2, -1, -1)$ of $X^3 + Y^3 + 7Z^3 = 0$, find 10 distinct solutions.

7

Non-singular cubics. The group law

Let $\mathcal{C}$ be a non-singular cubic curve and let **o** be a rational point on $\mathcal{C}$. We show that the set of rational points on $\mathcal{C}$ has a natural structure of commutative group with **o** as neutral element ("zero").

Here the ground field is arbitrary, the curve $\mathcal{C}$ is defined over it; and by rational point we mean point defined over the ground field.

The group law is defined as follows. Let **a**, **b** be rational points. Let **d** be the third point of intersection with $\mathcal{C}$ of the line through **a**, **b**. Let **e** be the third point of intersection of the line through **o**, **d**. Then we write

$$\mathbf{a} + \mathbf{b} = \mathbf{e}.$$

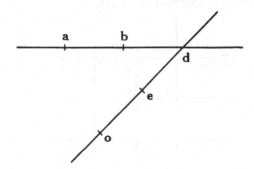

The construction has to be interpreted appropriately if two or more of the points involved coincide. For example if **b** = **a** we take the tangent at **a**.

We have to show that this operation "+" gives a structure of commutative group. Clearly

$$a + b = b + a$$

and

$$o + a = a$$

for all a.

Next we construct the inverse. Let the third intersection of the tangent at o be k. Let a^- be the third intersection of the line through a and k. Then by definition

$$a + a^- = o$$

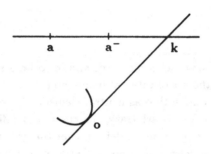

The crunch is to show that + is associative:

$$(a + b) + c = a + (b + c).$$

We give two proofs; the first geometric, the second more fundamental. Let a, b, c be given. Consider the diagram

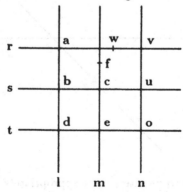

Here r, s, t, l, m, n are the names of lines and the remaining symbols

are points on C. All except $\mathbf{f}$, $\mathbf{w}$ are intersections of two of the lines. The whole figure is determined once $\mathbf{a}$, $\mathbf{b}$, $\mathbf{c}$ and $\mathbf{o}$ are given.

We have $(\mathbf{a}+\mathbf{b}) = \mathbf{e}$, and so $(\mathbf{a}+\mathbf{b})+\mathbf{c}$ is the third intersection of the line through $\mathbf{o}$, $\mathbf{f}$. Similarly $\mathbf{a} + (\mathbf{b} + \mathbf{c})$ is the third intersection of the line through $\mathbf{o}$, $\mathbf{w}$. To prove associativity, we thus have to show that $\mathbf{f}$, $\mathbf{w}$ are not as shown but coincide with the unlabelled intersection of the lines $\mathbf{r}$, $\mathbf{m}$.

We now recall a geometrical

Lemma 1. *Let* $\mathbf{x}_1, \dots, \mathbf{x}_8$ *be 8 points of the plane in general position*[4]. *Then there is a 9th point* $\mathbf{y}$ *such that every cubic curve through* $\mathbf{x}_1, \dots, \mathbf{x}_8$ *also passes through* $\mathbf{y}$.

We briefly recall the proof of the lemma. A cubic form $F(\mathbf{X})$, $\mathbf{X} = (X_1, X_2, X_3)$ has 10 coefficients. An equation $F(\mathbf{x}) = 0$ imposes a linear condition on the coefficients. Passing through $\mathbf{x}_1, \dots, \mathbf{x}_8$ imposes 8 conditions. Hence if $F_1(\mathbf{X})$, $F_2(\mathbf{X})$ are linearly independent forms through the 8 points, any other F is of the shape

$$F(\mathbf{X}) = \lambda F_1(\mathbf{X}) + \mu F_2(\mathbf{X}).$$

Now $F_1 = 0$, $F_2 = 0$ have 9 points in common; and clearly $F = 0$ passes through them all.

Now to the application of the Lemma. Let an equation for the line $\mathbf{l}$ be $l(\mathbf{X}) = 0$ etc. and consider the two (reducible) cubics

$$F_1(\mathbf{X}) = l(\mathbf{X})m(\mathbf{X})n(\mathbf{X}) = 0$$
$$F_2(\mathbf{X}) = r(\mathbf{X})s(\mathbf{X})t(\mathbf{X}) = 0.$$

Our nonsingular cubic C passes through 8 of the points of intersection of $F_1 = 0$, $F_2 = 0$ and so by the Lemma must pass through the 9th. Hence $\mathbf{f} = \mathbf{w}$, as required.

We now present a second proof of the associativity of the relation "+" for points which is more basic.

A linear form $l(\mathbf{X})$ (say) does not give a meaningful function on the curve C because the coefficients $\mathbf{X}$ are homogeneous. On the other hand, if $t(\mathbf{X})$ is another linear form, then the quotient

$$g(\mathbf{X}) = l(\mathbf{X})/t(\mathbf{X})$$

does give something meaningful. In the situation just discussed, the line

---

[4] This is the geometer's way of saying "such that the proffered proof works". In this case, what is needed is that the $\mathbf{x}_j$ give linearly independent conditions on the coefficients of F: so no 4 on a line and no 7 on a conic.

$l(\mathbf{X}) = 0$ passes through $\mathbf{a}$, $\mathbf{b}$, $\mathbf{d}$ and $t(\mathbf{X}) = 0$ through $\mathbf{d}$, $\mathbf{o}$, $\mathbf{e}$, all being points on $\mathcal{C}$. The function $g(\mathbf{X})$ thus has a zero $\mathbf{a}$, $\mathbf{b}$ and a pole at $\mathbf{o}$, $\mathbf{e}$. At the point $\mathbf{d}$ there is neither a zero nor a pole, as the zeros of the linear forms cancel out.

There is the notion of the order of a pole or zero at a nonsingular point of an algebraic curve which generalizes in an obvious way the notion of the order of a zero or pole of a rational function of a single variable. In our case, $g(\mathbf{X})$ clearly has simple zeros at $\mathbf{a}$, $\mathbf{b}$ and simple poles at $\mathbf{o}$, $\mathbf{e}$. The equation $\mathbf{e} = \mathbf{a} + \mathbf{b}$ is equivalent to the existence of such a function.

Similarly, the equation

$$\mathbf{x} = (\mathbf{a} + \mathbf{b}) + \mathbf{c}$$

is equivalent to the existence of a function with simple poles at $\mathbf{a}$, $\mathbf{b}$, $\mathbf{c}$, a double zero at $\mathbf{o}$ and a simple zero at $\mathbf{x}$. The equation

$$(\mathbf{a} + \mathbf{b}) + \mathbf{c} = \mathbf{a} + (\mathbf{b} + \mathbf{c})$$

is now obvious.

This point of view shows that the group law is unchanged under birational equivalence, since it depends only on the function field of the curve. The geometer would say that $\mathbf{a} + \mathbf{b} = \mathbf{c}$ precisely when the divisor $\{\mathbf{a}, \mathbf{b}\}$ is linearly equivalent to the divisor $\{\mathbf{o}, \mathbf{c}\}$.

We conclude with an informal explanation of what is meant by saying that a nonsingular cubic curve is of genus 1. Let $r \geq 2$ and let $\mathbf{x}_1, \cdots, \mathbf{x}_r, \mathbf{y}_1, \cdots, \mathbf{y}_{r-1}$ be points on $\mathcal{C}$, for simplicity all distinct. By manipulating linear forms in $\mathbf{X}$, as we did in the construction of $g(\mathbf{X})$, one can construct a function $h(\mathbf{X})$ on the curve where only poles are simple poles at $\mathbf{x}_1, \cdots, \mathbf{x}_r$ and which has zeros at $\mathbf{y}_1, \cdots, \mathbf{y}_{r-1}$. Then $h(\mathbf{X})$ has one further zero, which is completely determined.

Contrast the position on the line. Let $c_1, \cdots, c_r, d_1, \cdots, d_r$ be any $2r$ distinct numbers. Then the function

$$\prod_j (T - d_j) \Big/ \prod_j (T - c_j)$$

has simple zeros at the d_j, simple poles at the c_j and no further zeros or poles (even at infinity).

The genus of a curve is a measure of the freedom in imposing the zeros and poles of a function. The precise statement, which we shall not need, is slightly complicated and is called the Riemann-Roch Theorem.

§7. Exercises

1. Let $\mathbf{o}$, $\mathbf{a}$ be rational points on the nonsingular cubic $\mathcal{C}$. Construct the point $-\mathbf{a}$ with respect to the group law for which $\mathbf{o}$ is the neutral element.

2. Let $\mathbf{o}$, $\mathbf{o}_1$ be rational points on the nonsingular cubic $\mathcal{C}$. Show how the group law for which $\mathbf{o}_1$ is the neutral element can be expressed in terms of that for which $\mathbf{o}$ is the neutral element.

3. Let $\mathbf{o}$, $\mathbf{a}$ be rational points on the nonsingular cubic $\mathcal{C}$ and suppose that $3\mathbf{a} = \mathbf{o}$ with respect to the group law based on $\mathbf{o}$. Let $\mathbf{b} = 2\mathbf{a}$. Show that each side of the triangle $\mathbf{o}$, $\mathbf{a}$, $\mathbf{b}$ meets the tangent to $\mathcal{C}$ of the opposite vertex at a point of $\mathcal{C}$. Take $\mathbf{o}$, $\mathbf{a}$, $\mathbf{b}$ as the triangle of reference and express this condition in terms of the coefficients of the cubic form determining $\mathcal{C}$.

4. Let $\mathcal{C}$ be the curve
$$X^3 + Y^3 - XZ^2 - YZ^2 + 7XYZ = 0$$
and let $\mathbf{x} = (x,y,z)$ be a point on $\mathcal{C}$ defined over some $\mathbf{Q}_p$. Show that $y/x \to -1$ as $\mathbf{x} \to (0,0,1)$ (with respect to the p-adic topology).

5. In this question everything is defined over $\mathbf{Q}_p$ for some p. Let $\mathbf{a}$ be a nonsingular point on the cubic curve
$$F(X,Y,Z) = 0$$
and let $t(\mathbf{X}) = 0$ be the tangent. Let $l(\mathbf{X}) = 0$, $m(\mathbf{X}) = 0$ be lines through $\mathbf{a}$ distinct from the tangent. Show that there are d, e, f such that
$$dl(\mathbf{X}) + em(\mathbf{X}) + ft(\mathbf{X}) = 0$$
(identically) with $d \neq 0$, $e \neq 0$. Show that
$$m(\mathbf{x})/l(\mathbf{x}) \to -d/e$$
as $\mathbf{x} \to \mathbf{a}$.

8

Elliptic curves. Canonical Form

We are concerned with algebraic curves defined up to a birational equivalence over the ground field. For genus 0 we saw that every curve is equivalent to a conic (or line). For genus 1 no such reduction to a special form or forms is possible. The situation changes when we are also given a point on the curve which is defined over the ground field (a "rational point"). It is convenient to have a special name for this situation: an *elliptic curve* is a curve of genus 1 together with the specification of a rational point on it.

As canonical form we take

$$C: Y^2 = X^3 + AX + B$$

or, in homogeneous co-ordinates

$$Y^2 Z = X^3 + AXZ^2 + BZ^3.$$

The right hand side does not have multiple roots provided that

$$4A^3 + 27B^2 \neq 0.$$

The specified rational point o is the point $(X, Y, Z) = (0, 1, 0)$ at infinity.

Since the line at infinity is an inflexional tangent at o, the group law on C is especially simple:

$$-(x, y) = (x, -y)$$

and $a + b + c = o$ precisely when a, b, c are collinear.

We shall find this choice of canonical form particularly convenient when the ground field is $\mathbf{Q}$. When the ground field is of characteristic 2 or 3, we can no longer use C as a canonical form but must use

$$Y^2 + a_1 XY + a_3 Y = X^3 + a_2 X^2 + a_4 X + a_6.$$

However this is quite peripheral to our purposes and we leave it to the reader, if she wishes, to deal with these cases.

As we have not formally defined curves of genus 1, we will not give a formal proof that elliptic curves are birationally equivalent to the canonical form. In compensation we will give detailed algorithms for converting certain kinds of elliptic curves to that form. These could well be omitted at first reading.

Fact. (characteristic $\neq 2, 3$). *Any elliptic curve is birationally equivalent over the ground field to the canonical form for some A, B.*

More precisely the curve is equivalent to C and the equivalence takes the specified rational point **O** *on it into the point at infinity on C.*

Proof for the Cognoscenti. By the Riemann-Roch theorem, the set of functions on the curve with at worst a pole of order 2 at **O** has dimension 2. Let a basis be $1, \xi$. Similarly the set of functions with at worst a triple pole is of dimension 3 at **O**, with basis say $1, \xi, \eta$. Then the functions

$$\eta^2, \eta\xi, \eta, \xi^3, \xi^2, \xi, 1$$

all have at worst a pole of order 6. By the Riemann-Roch Theorem, there must be a linear relation between the 7 listed functions. The relation must involve both ξ^3 and η^2. A transformation

$$\xi \rightarrow c_1\xi + c_2$$
$$\eta \rightarrow c_3\eta + c_4\xi + c_5$$

reduces the relation to

$$\eta^2 = \xi^3 + A\xi + B$$

for some A, B.

Note for the Cognoscenti. The reason why there is no canonical form, or finite family of canonical forms for curves of genus 1 is that

$$2(g-1) = 0 \quad \text{for } g = 1.$$

For every other genus we can use the divisor of the differential of a function defined over the ground field to give a birational map. For example, for genus 2, there is always equivalence with some curve $Y^2 = \text{sextic in } X$.

Particular cases. The above proof does not, in any case, usually provide a practical algorithm. We discuss some special cases. Note that it is enough to transform the curve into the shape C. For if it takes $\mathbf{O}$ into $\mathbf{a}$, we can make the translation $\mathbf{x} \to \mathbf{x} - \mathbf{a}$ on C.

(i) *Cubic curve* $\mathcal{D}$. *Rational point* $\mathbf{O}$ *has inflexional tangent.* Here a linear tranformation of co-ordinates is enough, taking $\mathbf{O}$ to $\mathbf{o}$ and the tangent to be line at infinity.

For example

$$\mathcal{D} : X^3 + Y^3 + dZ^3 = 0$$

$$\mathbf{O} = (1, -1, 0).$$

Put

$$X = U + V, \qquad Y = U - V.$$

Then

$$6UV^2 = -2U^3 - dZ^3,$$

so

$$Y_1^2 Z_1 = X_1^3 - 2^4 . 3^3 . d^2 Z_1^3,$$

where

$$X_1 = -6dZ, \qquad Y_1 = 6^2 dV, \qquad Z_1 = U.$$

(ii) *Cubic curve* $\mathcal{D}$. *Rational point* $\mathbf{O}$ *not on inflexional tangent*[5].

The tangent at $\mathbf{O}$ meets $\mathcal{D}$ again at a rational point $\mathbf{P}$, say. We may take an affine system of co-ordinates with $\mathbf{P}$ as origin and with the tangent as Y-axis

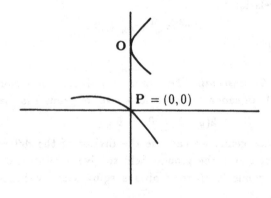

[5] The argument is due to Nagell: Sur les propriétés arithmétiques des cubiques planes du premier genre. *Acta Math.* **52** (1928-9), 92-106. Older geometrical techniques (adjoint curves etc.) had shown that every elliptic curve is birationally equivalent to a cubic, but he was the first to show that it can be reduced to the canonical form.

Then the curve $\mathcal{D}$ is given by $F(X, Y) = 0$, where
$$F(X, Y) = F_1(X, Y) + F_2(X, Y) + F_3(X, Y),$$
with F_j is homogeneous of degree j.

The Y-axis meets the curve at $(0, y)$, where
$$0 = y F_1(0, 1) + y^2 F_2(0, 1) + y^3 F_3(0, 1).$$

Since the Y-axis is a tangent, we have a double root:
$$F_2(0, 1)^2 - 4F_1(0, 1)F_3(0, 1) = 0. \qquad (*)$$

Now consider the intersection of the curve with $Y = tX$. Then
$$0 = x F_1(1, t) + x^2 F_2(1, t) + x^3 F_3(1, t).$$

Discarding the solution $x = 0$, we have
$$s^2 = F_2(1, t)^2 - 4F_1(1, t)F_3(1, t)$$
$$= G(t) \qquad \text{(say)},$$
where
$$s = 2F_3(1, t)x + F_2(1, t).$$

Now $G(t)$ is a cubic by $(*)$; and we achieve the canonical form by a linear transformation on s, t.

(iii) *Curve $\mathcal{D}$ is $Y^2 = $ Quartic in X with rational point.*

Let the rational point be (a, b). By a transformation
$$X \to \frac{1}{X - a}, \qquad Y \to \frac{Y}{(X - a)^2},$$
we may suppose that the rational point is at infinity:
$$Y^2 = f_0 + f_1 X + f_2 X^2 + f_3 X^3 + f_4 X^4,$$
where f_4 is a square. On dividing by f_4, we have without loss of generality
$$f_4 = 1.$$

We can write the right hand side as
$$G(X)^2 + H(X),$$
where
$$G(X) = X^2 + g_1 X + g_0$$
$$H(X) = h_1 X + h_0,$$
and the g_j, h_j are easily given in terms of the f_j.

The equation of the curve is now
$$(Y + G(X))(Y - G(X)) = H(X).$$

Put
$$Y + G(X) = T,$$

so

$$Y - G(X) = \frac{H(X)}{T}$$

and

$$2G(X) = T - \frac{H(X)}{T}.$$

Multiply by T^2 and put $TX = S$. We get

$$2S^2 + 2g_1 TS + 2g_0 T^2 = T^3 - h_1 S - h_0 T.$$

This is readily brought to the canonical form.

(iv) *Intersection of two quadric surfaces with a rational point.*

We use homogeneous co-ordinates X, Y, Z, T and may suppose that the common rational point is $(0, 0, 0, 1)$. The two quadric forms are thus of the shape

$$\left. \begin{array}{l} Q_1 = TL + R \\ Q_2 = TM + S \end{array} \right\},$$

where L, M are linear in X, Y, Z and R, S are quadratic.

Suppose, first, that L and M are linearly dependent. Then without loss of generality $M = 0$. The intersection is

$$S(X, Y, Z) = 0, \qquad T = R(X, Y, Z)/L(X, Y, Z);$$

which is of genus 0.

Otherwise, eliminating T, we have

$$C(X, Y, Z) = LS - RM = 0,$$

where C is a homogeneous cubic. It has the rational point

$$L(X, Y, Z) = M(X, Y, Z) = 0.$$

Hence we are reduced to an earlier case.

§8. Exercises

1. Transform the following curves to canonical form:

(i) $X^3 + Y^3 + dZ^3 = 0$
(ii) $X^3 + Y^3 + Z^3 - 3mXYZ = 0$
(iii) $Y^2 - kT^2 = X^2$, $Y^2 + kT^2 = Z^2$
(iv) $X_1^2 X_2 - X_1 X_2^2 - X_1 X_3^2 + X_2^2 X_3 = 0$

2. [Difficult]. Show that the group law on

$$X^2 = Y^2 - T^2, \qquad Z^2 = Y^2 + T^2$$

with $(1, 1, 1, 0)$ as neutral element is given by $x_3 = x_1 + x_2$, where

$$x_3 = x_2 t_2 y_1 z_1 - x_1 t_1 y_2 z_2$$
$$y_3 = y_2 t_2 z_1 x_1 - y_1 t_1 z_2 x_2$$
$$z_3 = z_2 t_2 x_1 y_1 - z_1 t_1 x_2 y_2$$
$$t_3 = t_2^2 x_1^2 - t_1^2 x_2^2 = t_2^2 y_1^2 - t_1^2 y_2^2 = t_2^2 z_1^2 - t_1^2 z_2^2.$$

3. (i) Find all the points defined over the field $\mathbf{F}_5$ of 5 elements on each of

$$Y^2 Z = X^3 + X Z^2$$
$$Y^2 Z = X^3 + 2X Z^2$$
$$Y^2 Z = X^3 + Z^3.$$

Check in each case that they form a group under the group law, with $(0, 1, 0)$ as neutral element.

(ii)   As (i) but with other $\mathbf{F}_p$ and other curves

$$Y^2 Z = X^3 + AX Z^2 + B Z^3.$$

Find an example where the group is not cyclic. Can you find an example where the group requires more than 2 generators?

4. In the curves considered below, the point at infinity is taken as neutral element for the group law.

(i)   Let $Y^2 = (X - \alpha)(X^2 + aX + b)$ be an elliptic curve. Show that the transformation $x \to x + (\alpha, 0)$ induces a fractional-linear transformation

$$T : x \to (t_{11} x + t_{12})/(t_{21} x + t_{22}).$$

Check that $T^2 : x \to x$.

(ii)   Consider $Y^2 = (X - \alpha_1)(X - \alpha_2)(X - \alpha_3)$ and let T_1, T_2, T_3 be as in (i) with $\alpha = \alpha_j$ $(j = 1, 2, 3)$. Show that T_1, T_2, T_3 commute and that

$$T_1 T_2 T_3 : x \to x.$$

(iii)   Let T_j be the 2×2 matrix of coefficients $\left(\begin{smallmatrix} t_{11} & t_{21} \\ t_{12} & t_{22} \end{smallmatrix} \right)$ in (i) with $\alpha = \alpha_j$ $(j = 1, 2, 3)$. Show that

$$T_1 T_2 + T_2 T_1 = 0.$$

(iv)   Find the fixed points of T_1 and show that they are interchanged by T_2.

5. Find a necessary and sufficient condition that a line $Y = lX + m$

should be an inflexional tangent to

$$Y^2 = X^3 + AX + B.$$

Hence find a general formula for the curves in canonical form having a rational point of order 3.

6. Find a necessary and sufficient condition that a line $Y = lX + m$ should be an inflexional tangent to $Y^2 = X(X^2 + aX + b)$.

Hence find a general formula for curves in canonical form having a point of order 6.

7. Let

$$F(X, Y, Z) = X^2Y + XZ^2 + 2Y^3 + Z^3.$$

Find a birational transformation defined over $\mathbf{Q}$ taking the curve $F = 0$ into canonical form with the point $(1, 0, 0)$ going to the point at infinity.

8. Find a birational transformation defined over $\mathbf{Q}$ taking

$$X_1^2 - 2X_2^2 + X_3^2 = 0, \quad X_2^2 - 2X_3^2 + X_4^2 = 0$$

into canonical form, with $(1, 1, 1, 1)$ going to the point at infinity.

9. Invent similar exercises to the two preceding, and solve them.

9

Degenerate laws

In this section we consider the curve
$$C : Y^2 = X^3 + AX + B \qquad (1)$$
when
$$4A^3 + 27B^2 = 0. \qquad (2)$$
There is then precisely one singular point. We recall that if (2) does not hold, there is a group law on the curve given by[6]
$$\mathbf{a} + \mathbf{b} + \mathbf{c} = 0$$
whenever $\mathbf{a}$, $\mathbf{b}$, $\mathbf{c}$ are the intersection of a line with C. We show that this continues to give a group law on the nonsingular points in the degenerate case (2), and we find out what it is.

There are two cases, the second with two subcases.

First case. Cusp. Suppose $A = B = 0$, so
$$C : Y^2 Z = X^3$$
with a singular point at the origin. Any line not passing through the origin can be written
$$Z = lX + mY.$$
It meets C where
$$X^3 - Y^2(lX + mY) = 0$$

---

[6] We write indifferently 0 or o for the neutral element of the group law.

If the three points of intersection are (x_j, y_j, z_j) $(j = 1, 2, 3)$, it follows that

$$u_1 + u_2 + u_3 = 0,$$

where

$$u_j = x_j / y_j.$$

We therefore have the additive group, the zero being the point $(0, 1, 0)$ at infinity.

Second case[7]. *Double point.* (Characteristic $\neq 2$). If not both A, B vanish, then, after a transformation $X \rightarrow X +$ constant, we have

$$\mathcal{C} : Y^2 Z = X^2 (X + CZ) \qquad (C \neq 0),$$

i.e.

$$(Y^2 - CX^2) Z = X^3.$$

Suppose, first, that $C = \gamma^2$ is a square. Put

$$U = Y + \gamma X, \qquad V = Y - \gamma X;$$

so $\mathcal{C}$ is given by

$$8\gamma^3 U V Z = (U - V)^2.$$

Any line not passing through the origin can be written

$$Z = lU + mV.$$

It meets $\mathcal{C}$ where

$$(U - V)^3 - 8\gamma^3 U V (lU + mV) = 0.$$

If the points of intersection are (u_j, v_j, z_j) $(j = 1, 2, 3)$, then

$$\left(\frac{u_1}{v_1} \right) \left(\frac{u_2}{v_2} \right) \left(\frac{u_3}{v_3} \right) = 1.$$

We have the multiplicative group.

Now suppose that C is not a square. Adjoin γ to the ground field, where $\gamma^2 = C$. For a point (x, y, z) on $\mathcal{C}$, put

$$\frac{y + \gamma x}{y - \gamma x} = r + s\gamma \qquad \text{(say)},$$

where

$$r^2 - s^2 C = 1. \qquad (*)$$

We now have a "twisted" multiplication law on (*). Compare the multiplication of the complex numbers $x + iy$ with $x^2 + y^2 = 1$.

---

[7] We shall not require the details about this case in later work.

Note for the Cognoscenti. In characteristic 2 the curve

$$\mathcal{C}: \quad Y^2 Z = X^3 + AXZ^2 + BZ^3$$

is always singular. Write the equation as

$$(Y^2 - BZ^2)Z = X(X^2 + AZ^2).$$

Over a finite (or, more generally, a perfect) field, we have

$$B = \beta^2, \qquad A = \alpha^2$$

for some α, β. Then the curve is

$$(Y + \beta Z)^2 Z = X(X + \alpha Z)^2;$$

which is clearly singular.

If the ground field is not perfect, we may have an example of a singularity defined over an inseparable extension, compare footnote in §6.

10

Reduction

The philosophy is to approach the rational field $\mathbf{Q}$ through the local fields $\mathbf{Q}_p$ and, similarly, to approach the $\mathbf{Q}_p$ through the finite fields $\mathbf{F}_p$ by reduction modulo p. We do no more than is required for the applications.

The mod p map $\mathbf{Z}_p \to \mathbf{F}_p$ is denoted by a bar $a \to \overline{a}$. This is extended to the corresponding 2-dimensional projective planes V, $\overline{V}$ as follows. Let (a_1, a_2, a_3) be projective co-ordinates of a point $\mathbf{a}$ of V. By multiplying a_1, a_2, a_3 by the same element of $\mathbf{Q}_p$, we have without loss of generality

$$\max\{|a_1|, |a_2|, |a_3|\} = 1,$$

where $\| = \|_p$. Then $(\overline{a_1}, \overline{a_2}, \overline{a_3})$ are the co-ordinates of a well-defined point $\overline{\mathbf{a}}$ of $\overline{V}$.

In a similar way, we define the reduction $\overline{\mathbf{l}}$ of a line

$$\mathbf{l}: \quad l_1 X_1 + l_2 X_2 + l_3 X_3 = 0.$$

If the point $\mathbf{a}$ lies on the line $\mathbf{l}$, then clearly $\overline{\mathbf{a}}$ lies on $\overline{\mathbf{l}}$.

We need only the least sophisticated of the many ways of reducing a cubic curve

$$\mathcal{C}: \quad F(\mathbf{X}) = 0$$

defined over $\mathbf{Q}_p$. Here

$$F(\mathbf{X}) = \sum_{i \le j \le k} f_{ijk} X_i X_j X_k \in \mathbf{Q}_p[\mathbf{X}]$$

where the $f_{ijk} \in \mathbf{Q}_p$ are not all 0 and without loss of generality

$$\max_{i,j,k} |f_{ijk}| = 1.$$

Then

$$\overline{F}(\mathbf{X}) = \sum_{i \le j \le k} \overline{f}_{ijk} X_i X_j X_k| \in \mathbf{F}_p[\mathbf{X}]$$

is not the zero polynomial, and defines the reduced curve

$$\overline{C} : \ \overline{F}(\mathbf{X}) = 0$$

over $\mathbf{F}_p$. It may, of course, be reducible[8].

If a point **a** lies on C, then clearly $\overline{\mathbf{a}}$ lies on $\overline{C}$. There is a weak converse

Lemma 1. *Let* $\overline{\mathbf{b}}$ *be a nonsingular point of* $\overline{C}$. *Then there is an* **a** *on* C *such that* $\overline{\mathbf{a}} = \overline{\mathbf{b}}$.

Note. The notation $\overline{\mathbf{b}}$ is intended to denote a point defined over $\mathbf{F}_p$ not necessarily derived from a **b**. We say that $\overline{\mathbf{b}}$ *lifts* to **a**. It is easy to see by examples that a singular point on $\overline{C}$ may or may not lift to a point of C (cf. Exercises).

We construct **a** by successive approximation à la Newton. The generic term for such constructions in p-adic analysis is Hensel's Lemma.

Lemma 2. *Let* $G(T) \in \mathbf{Z}_p[T]$ *and let* $t_0 \in \mathbf{Z}_p$ *be such that*

$$|G(t_0)| < 1, \qquad |G'(t_0)| = 1,$$

where G' *is the formal derivative of* G. *Then there is a* $t \in \mathbf{Z}_p$ *such that*

$$G(t) = 0 \qquad |t - t_0| \le G(t_0).$$

Assuming the truth of Hensel's Lemma for the moment, we complete the proof of the Lemma. Since $\overline{\mathbf{b}}$ is nonsingular on $\overline{C}$, we may suppose that

$$\frac{\partial \overline{F}}{\partial X_1}(\overline{\mathbf{b}}) \ne 0.$$

Pick any $b_j \in \mathbf{Z}_p$ such that $\overline{\mathbf{b}} = (\overline{b}_1, \dots, \overline{b}_n)$. Then the conditions of Hensel's Lemma apply to

$$G(T) = F(T, b_2, \dots, b_n), \qquad t_0 = b_1.$$

Put $\mathbf{a} = (t, b_2, \dots, b_n)$, where t is provided by Hensel. Clearly $F(\mathbf{a}) = 0$, $\overline{\mathbf{a}} = \overline{\mathbf{b}}$, so **a** does what is required.

It remains to prove the Hensel's Lemma. Let U be an indeterminate.

---

[8] In the sense that $\overline{F}(\mathbf{X})$ factorizes. There is an unfortunate clash of meanings between "reduced" (mod p) and "reducible".

Then

$$G(T + U) = G(T) + UG_1(T) + U^2 G_2(T) + \dots$$

where $G_j \in \mathbf{Z}_p[T]$ and $G_1 = G'$. Now define

$$u = -G(t_0)/G'(t_0),$$

so

$$G(t_0 + u) = u^2 G_2(t_0) + u^3 G_3(t_0) + \dots .$$

Hence

$$|G(t_1)| \leq |G(t_0)|^2,$$

where

$$t_1 = t_0 + u.$$

Clearly

$$|G'(t_1)| = |G'(t_0)| = 1.$$

We may therefore iterate the process and get a fundamental sequence t_j $(t \geq 0)$. The limit t clearly does what is required.

We shall also need information about the behaviour of the intersection of a line and a cubic curve under reduction. From what we have already proved, if l meets C in a, then $\bar{l}$ meets $\overline{C}$ in $\bar{a}$. But suppose that l meets C in a, b with a $\neq$ b: if $\bar{a} = \bar{b}$, can we be sure that it has multiplicity ≥ 2 in the intersection?

The following lemma confirms expectations.

Lemma 3. *Suppose that the line* l *meets the cubic curve* C *in* a, b, c, *multiple points of intersection being given with their multiplicities. Then either*

(I)   *the entire line* $\bar{l}$ *is in* $\overline{C}$ *or*

(II) $\bar{l}$ *meets* $\overline{C}$ *in* $\bar{a}$, $\bar{b}$, $\bar{c}$, *multiple points occuring with the correct multiplicities.*

Proof. We have without loss of generality

$$l_3 = 1 = \max(|l_1|, |l_2|, |l_3|).$$

Consider

$$G(X_1, X_2) = F(X_1, X_2, -l_1 X_1 - l_2 X_2)$$
$$= \mathbf{Z}_p[X_1, X_2].$$

Its reduction is

$$\overline{G}(X_1, X_2) = \overline{F}(X_1, X_2 - \bar{l}_1 X_1 - \bar{l}_2 X_2).$$

If $\overline{G}(X_1, X_2) = 0$, we have case (I) of the Lemma, so we may suppose that

$$\overline{G}(X_1, X_2) \neq 0.$$

We normalize the coefficients of **a**, **b**, **c** so that

$$\max(|a_1|, |a_2|, |a_3|) = 1.$$

Since $\mathbf{la} = 0$, it follows that

$$(\overline{a}_1, \overline{a}_2) \neq (0, 0)$$

etc.

By hypothesis, there is some $\lambda \in \mathbf{Q}_p$ such that

$$G(X_1, X_2) = \lambda(a_2 X_1 - a_1 X_2)(b_2 X_1 - b_1 X_2)(c_2 X_1 - c_1 X_2)$$
$$= \lambda H(X_1, X_2).$$

Now

$$\overline{H}(X_1, X_2) = (\overline{a}_2 X_1 - \overline{a}_1 X_2)(\overline{b}_2 X_1 - \overline{b}_1 X_2)(\overline{c}_2 X_1 - \overline{c}_1 X_2)$$
$$\neq 0.$$

Hence $\overline{G}$, $\overline{H}$ differ only by a scalar multiple, which is what we needed to prove.

§10. Exercises

1. (i) Let $\mathcal{C}$ be the curve $Y^2 = X^3 + p$ over $\mathbf{Q}_p$. Show that the point $(0, 0)$ on the mod p curve does not lift to a point of $\mathcal{C}$.

(ii) Find an example of an elliptic curve $\mathcal{C}$ over $\mathbf{Q}_p$ such that the mod p curve has a cusp which is the reduction of a point on $\mathcal{C}$.

2. Find examples of curves $\mathcal{C}$ over $\mathbf{Q}_p$ such that the mod p curve has a double point with distinct tangents which (i) lifts, (ii) does not lift, to $\mathcal{C}$.

11

The p-adic case

Let
$$\mathcal{C}: \; Y^2 = X^3 + AX + B$$
be an elliptic curve defined over $\mathbf{Q}_p$, so
$$4A^3 + 27B^2 \neq 0$$
and, without loss of generality,
$$A, B \in \mathbf{Z}_p.$$
In this section we study the group $\mathfrak{G}$ of points on $\mathcal{C}$ defined over $\mathbf{Q}_p$.

Our tool will be the theory of reduction developed in the preceeding section. For this, we write $\mathcal{C}$ homogeneously
$$\mathcal{C}: \; Y^2Z = X^3 + AXZ^2 + BZ^3.$$
The reduced curve
$$\overline{\mathcal{C}}: \; Y^2Z = X^3 + \overline{A}XZ^2 + \overline{B}Z^3$$
over $\mathbf{F}_p$ may be singular but (with an eye to Lemma 3 of §10) we note that $\overline{\mathcal{C}}$ does not contain a line.

Let $\overline{\mathfrak{G}}$ denote the set of points on $\overline{\mathcal{C}}$ defined over $\mathbf{F}_p$ and let $\overline{\mathfrak{G}}^{(0)} \subset \overline{\mathfrak{G}}$ be the non-singular points. Write $\mathfrak{G}^{(0)} \subset \mathfrak{G}$ for the set of points which reduce mod p to $\overline{\mathfrak{G}}^{(0)}$. The map
$$\mathfrak{G}^{(0)} \to \overline{\mathfrak{G}}^{(0)}$$
is surjective by Lemma 1 of §10.

How does the group structure behave? Let $\mathbf{a}, \mathbf{b}, \mathbf{c} \in \mathfrak{G}$ with
$$\mathbf{a} + \mathbf{b} + \mathbf{c} = \mathbf{o}.$$

This holds if and only if **a**, **b**, **c** are the intersection of C with a line l. Then the reductions $\overline{\mathbf{a}}$, $\overline{\mathbf{b}}$, $\overline{\mathbf{c}}$ are the intersections of $\overline{C}$ with $\overline{\mathbf{l}}$. On $\overline{C}$ we have defined a group law only for the non-singular points. If $\overline{\mathbf{a}}$, $\overline{\mathbf{b}}$, $\overline{\mathbf{c}} \in \overline{\mathfrak{G}}^{(0)}$, then

$$\overline{\mathbf{a}} + \overline{\mathbf{b}} + \overline{\mathbf{c}} = \overline{\mathbf{o}}.$$

To sum up so far, we have a subgroup $\mathfrak{G}^{(0)}$ of $\mathfrak{G}$ such that there is a group homomorphism $\mathfrak{G}^{(0)} \to \overline{\mathfrak{G}}^{(0)}$ onto $\overline{\mathfrak{G}}^{(0)}$. The kernel of this homomorphism is the set of points which map into $\overline{\mathbf{o}}$, that is, in inhomogeneous co-ordinates, **o** itself together with the $(x, y) \in \mathfrak{G}$ with $x \notin \mathbf{Z}_p$, $y \notin \mathbf{Z}_p$. This is called the *kernel of the reduction.*

Next, we look at the structure of the kernel of reduction. If $(x, y) \in \mathfrak{G}$, $x, y \notin \mathbf{Z}_p$, then clearly $|y|^2 = |x|^3$ and so

$$|x| = p^{2n}, \qquad |y| = p^{3n}$$

for some $n \geq 1$. We call n the *level* of (x, y). For (x, y) not in the kernel of reduction the level is 0, by definition. The level of **o** is ∞.

Now for integer $N \geq 1$ make the transformation

$$X_N = p^{2N} X, \qquad Y_N = p^{3N} Y, \qquad Z_N = Z,$$

so the equation of C becomes

$$C_N : \quad Y_N^2 Z_N = X_N^3 + p^4 A X_N Z_N^2 + p^6 B Z_N^3.$$

We may use the new co-ordinates for a reduction mod p: the reduced curve is

$$\overline{C}_N : \quad Y_N^2 Z_N = X_N^3.$$

We can now transfer what was done earlier to the new situation. A point (x, y) maps into the singular point $(0, 0)$ of $\overline{C}_N$ if its level is $< N$. It is in the kernel of reduction for C_N if its level is $> N$. Finally, the group of the non-singular points on the $\overline{C}_N$ defined over $\mathbf{F}_p$ is the additive group of $\mathbf{F}_p$. They are in the image of $\mathfrak{G}$, as before.

For $N \geq 1$ define $\mathfrak{G}^{(N)}$ to be the set of points of $\mathfrak{G}$ of level $\geq N$. We have proved

Lemma 1. *The $\mathfrak{G}^{(N)}$ are groups and*

$$\mathfrak{G} \supset \mathfrak{G}^{(0)} \supset \mathfrak{G}^{(1)} \supset \cdots \supset \mathfrak{G}^{(N)} \supset \cdots.$$

The quotient graphs of $\mathfrak{G}^{(N)}/\mathfrak{G}^{(N+1)}$ for $N \geq 1$ are cyclic of order p. The quotient $\mathfrak{G}^{(0)}/\mathfrak{G}^{(1)}$ is isomorphic to the group of nonsingular points on $\overline{C}$. Further,

$$\bigcap_N \mathfrak{G}^{(N)} = \{\mathbf{o}\}.$$

The sequence of groups is called the *p-adic filtration*.

Corollary. *Let* $\mathbf{x} = (x, y) \in \mathfrak{G}$ *be of finite order prime to* p. *Then* x, $y \in \mathbf{Z}_p$.

Proof. Otherwise $\mathbf{x}$ is of some level $n \geq 1$. Then $\mathbf{x} \in \mathfrak{G}^{(n)}$, $\mathbf{x} \notin \mathfrak{G}^{(n+1)}$ and so maps into a non-zero element of $\mathfrak{G}^{(n)}/\mathfrak{G}^{(n+1)}$. But this is of order p.

  Our next aim is to free the statement in the Corollary from the requirement that the order is prime to p.

  The homomorphism of $\mathfrak{G}^{(N)}/\mathfrak{G}^{(N+1)}$ to the additive group mod p is given by

$$(x, y) \to p^{-N} x/y \mod p.$$

For $\mathbf{x} \in \mathfrak{G}^{(1)}$ we introduce $u(\mathbf{x})$ defined by

$$u(\mathbf{x}) = x/y \qquad (\mathbf{x} = (x, y)),$$
$$u(\mathbf{o}) = 0.$$

Note that $|u(\mathbf{x})| = p^{-n}$, where n is the level of $\mathbf{x}$.

Lemma 2. *Let* $\mathbf{x}_1$, $\mathbf{x}_2 \in \mathfrak{G}^{(1)}$. *Then*

$$|u(\mathbf{x}_1 + \mathbf{x}_2) - u(\mathbf{x}_1) - u(\mathbf{x}_2)| \leq \max\{|u(\mathbf{x}_1)|^5,\ |(\mathbf{x}_2)|^5\}.$$

Proof. We may suppose that none of $\mathbf{x}_1$, $\mathbf{x}_2$, $\mathbf{x}_1 + \mathbf{x}_2$ is $\mathbf{o}$. Without loss of generality

$$|u(\mathbf{x}_1)| \geq |u(\mathbf{x}_2)|.$$

Define N to be the level of $\mathbf{x}_1$. We use the co-ordinates X_N, Y_N and the curve $\mathcal{C}_N$ introduced above.

  Since neither $\mathbf{x}_1$, nor $\mathbf{x}_2$ maps into the singularity $(0, 0)$ of $\overline{\mathcal{C}}_N$, the line joining them has the shape

$$Z_N = lX_N + mY_N,$$

where

$$|l| \leq 1, \qquad |m| \leq 1.$$

This meets $\mathcal{C}$ where

$$\begin{aligned}
0 = &- Y_N^2(lX_N + mY_N) + X_N^3 \\
&+ p^{4N} A X_N(lX_N + mY_N)^2 \\
&+ p^{6N} B(lX_N + mY_N)^3 \\
= &\, c_3 X_N^3 + c_2 X_N^2 Y_N + c_1 X_N Y_N^2 + c_0 Y_N^3 \qquad (*)
\end{aligned}$$

(say). Here

$$c_3 = 1 + p^{4N} A l^2 + p^{6N} B l^3$$
$$c_2 = 2p^{4N} lmA + 3p^{6N} l^2 mB,$$

so

$$|c_3| = 1, \qquad |c_2| \le p^{-4N}.$$

The roots X_N/Y_N of (*) are $-p^{-N}u(\mathbf{x}_1 + \mathbf{x}_2)$, $p^{-N}u(\mathbf{x}_1)$ and $p^{-N}u(\mathbf{x}_2)$. Since the sum of the roots is $-c_2/c_3$, the result follows.

Corollary 1.

$$|u(s\mathbf{x})| = |s| \, |u(\mathbf{x})|$$

for all $\mathbf{x} \in \mathfrak{G}^{(1)}$ *and all* $s \in \mathbf{Z}$.

Proof. By induction, for $s > 0$ we have

$$|u(s\mathbf{x}) - su(\mathbf{x})| \le |u(\mathbf{x})|^5.$$

This proves the result for $p \nmid s$ and for $s = p$. Now use induction on the power of p in s.

Corollary 2. $\mathfrak{G}^{(1)}$ *is torsion-free.*

Corollary 3. *Suppose that* $p \ne 2$, $|4A^3 + 27B^2| = 1$. *Then the torsion subgroup of* $\mathfrak{G}$ *is isomorphic to a subgroup of* $\overline{\mathfrak{G}}$.

Proof. For $\mathfrak{G} = \mathfrak{G}^{(0)}$; and so

$$\overline{\mathfrak{G}} = \mathfrak{G}/\mathfrak{G}^{(1)},$$

where $\mathfrak{G}^{(1)}$ is torsion free.

Note for the Cognoscenti. This all generalizes to algebraic extensions of $\mathbf{Q}_p$. The proof that torsion points of order prime to p have integral co-ordinates continues to hold, but that for points of p-power order may break down if there is ramification.

There is a power-series in $u = u(\mathbf{x})$ which gives a parametrization of the group $\mathfrak{G}^{(N)}$ for large enough N. This was originally shown by transferring the formulae from the complex variable case. A modern approach is by formal groups and formal logarithms, see, for example, Silverman's book.

12

Global torsion

Let
$$C: \ Y^2 = X^3 + AX + B$$
be an elliptic curve over $\mathbf{Q}$, so
$$4A^3 + 27B^2 \neq 0$$
and without loss of generality
$$A, B \in \mathbf{Z}.$$

Theorem 1. *The group of rational points on C of finite order is finite. If $(x, y) \neq \mathbf{o}$ is of finite order, then*
$$x, \ y \in \mathbf{Z}$$
and
$$y = 0 \quad or \quad y^2 \mid (4A^3 + 27B^2).$$

Proof. Let $\mathfrak{G}$ be the group of points on C defined over $\mathbf{Q}$ and let $\mathfrak{G}_p$ be the group for $\mathbf{Q}_p$, where p runs through the primes.

Let $(x, y) \neq \mathbf{o}$ be torsion. Since $\mathfrak{G} \subset \mathfrak{G}_p$ we have
$$x \in \mathbf{Z}_p, \qquad y \in \mathbf{Z}_p$$
for all p, and so
$$x \in \mathbf{Z}, \qquad y \in \mathbf{Z}.$$

Now let p be any prime with $p \neq 2$, $p \nmid (4A^3 + 27B^2)$. Then by the last Corollary of §11, the torsion group of $\mathfrak{G}$ is isomorphic to a subgroup of the group of points over $\mathbf{F}_p = \mathbf{Z}$ mod p. Hence the torsion group is finite. By looking at different p, one can in general restrict the order of the torsion group severely. But the following argument makes it easy to find the torsion points themselves.

If $2(x, y) = o$, then $y = 0$. Otherwise, $2(x, y) = (x_2, y_2)$ (say) is also torsion, so x_2, $y_2 \in \mathbf{Z}$.

Now taking the tangent at (x, y), we have (cf. Formulary)

$$x_2 + 2x = \left(\frac{3x^2 + A}{2y}\right)^2 = \frac{(3x^2 + A)^2}{4(x^3 + Ax + B)}.$$

and so $y^2 = x^3 + Ax + B$ divides $(3x^2 + A)^2$.

But now,

$$(3X^2 + 4A)(3X^2 + A)^2 \equiv 4A^3 + 27B^2$$
$$\mathrm{mod}\,(X^3 + AX + B) \qquad (*)$$

in $\mathbf{Z}[X, A, B]$, as in readily verified. Hence

$$y^2 \mid (4A^3 + 27B^2),$$

as required. [For more on identity $(*)$, see §16].

Note. There are stronger statements about the torsion of $\mathcal{C}$ when $AB = 0$, see Exercises. Mazur has determined all possible forms of the torsion group. It is one of

$$\mathbf{Z}/n\mathbf{Z} \qquad 1 \le n \le 10 \qquad \text{or} \qquad n = 12$$

or

$$\mathbf{Z}/2\mathbf{Z} \times \mathbf{Z}/2n\mathbf{Z} \qquad 1 \le n \le 4;$$

all of which occur.

§12. Exercises

1. Find the torsion groups over $\mathbf{Q}$ of the following elliptic curves:

(i) $Y^2 = X^3 + 1$
(ii) $Y^2 = X^3 - 43X + 166$
(iii) $Y^2 = X^3 - 219X + 1654$
(iv) $Y^2 = X(X - 1)(X + 2)$
(v) $Y^2 = X(X + 1)(X + 4)$
(vi) $X^3 + Y^3 + Z^3 - 15XYZ = 0$
(vii) $Y^2 = X(X + 81)(X + 256)$
(viii) $X_1^2 X_2 - X_1 X_2^2 - X_1 X_3^2 + X_2^2 X_3 = 0$
[*Note*: not a random sample!]

2. Fill in the details of the sketched proof of the following theorem [9] [or find a better one!].

Theorem. *Let $A \in \mathbf{Z}$ be 4-th power free. Then all the torsion points on*

$$C : Y^2 = X(X^2 + A)$$

are given by (I), (II), (III) below:

(I) (0,0) *of order 2.*

(II) *If $A = 4$, the points $(2, \pm 4, 1)$ of order 4.*

(III) *If $A = -C^2$, $C \in \mathbf{Z}$, the points $(\pm C, 0)$ of order 2.*

Sketch proof.

(i) If $(x, y) = 2(a, b)$, then

$$x = (a^2 - A)^2 / 4b^2.$$

(ii) The points of order 2 are as stated.

(iii) $(0,0) = 2(a, b)$ for some (a, b) precisely when $A = 4$. The $(\pm C, 0)$ are never of form $2(a, b)$.

From now on, let (a, b) be a point of *odd* order.

(iv) $a = \square$

(v) If $d = \gcd(a, A)$ then $a = da_1$, $A = dA_1$ $b = db_1$ where $b_1^2 = a_1(da_1^2 + A_1)$.

(vi) There exists f, g, h such that $\gcd(f, g) = 1$ and $a_1 = \pm f^2$, $da_1^2 + A_1 = \pm g^2$, $b_1 = fg$, $d = \pm h^2$

(vii) $a^2 - A = 2h^4 f^4 - h^2 g^2$, $b = \pm h^2 fg$.

(viii) $a^2 - A \equiv 0 \pmod{2b}$.

(ix) Hence $f = 1$, $g \equiv 0$ (2), $h \equiv 0$ (2). [*Hint.* First show that $f \mid g$].

(x) Hence $2^4 \mid A$.

(xi) Contradiction!

3. Fill in the sketched proof of the following theorem [10] [or find a better].

Theorem. *Let $B \in \mathbf{Z}$ be 6-th power free and let*

$$C : \quad Y^2 = X^3 + B.$$

All the torsion points are given by the following.

(I) *If $B = C^2$, the points $(0, \pm C)$ of order 3.*

(II) *If $B = D^3$, the points $(-D, 0)$ of order 2.*

[9] cf. T.Nagell. Solution de quelques problèmes dans la théorie arithmétique des cubiques planes du premier genre. *Skrifter utg. av det norske vidensk.-akad i Oslo, Mat.-naturv. kl.* **1935**, No 1, 1-25.

[10] The result is due to R.Fueter: Ueber kubische diophantische Gleichungen. *Comm. Math. Helv.* **2** (1930), 68-89; but the argument suggested is based loosely on L.J.Mordell. The infinity of rational solutions of $y^2 = x^3 + k$. *J. London Math. Soc.* **41** (1966), 523-525.

(III) *If $B = 1$, the points $(2, \pm 3)$ of order 6.*
(IV) *If $B = -432 = -2^4.3^3$, the points $(12, \pm 36)$ of order 3.*

Sketch proof.

(i)   If $(x, y) = 2(a, b)$, $b \neq 0$ then
$$x = (w - 2)a, \qquad w = 9a^3/4b^2.$$

(ii) the elements of 2-power order are as stated.

(iii) Elements $(0, b)$ are of order 3.

From now on, let (a, b) be of odd order with $a \neq 0$. The strategy is to show that $w \in \mathbf{Z}$. The cases with $w = 1, 2, 3$ are then easily dealt with. Otherwise, $|x|_\infty > |a|_\infty$ and so on repeated duplication $|x|_\infty \to \infty$ a contradiction. We sketch a proof that $w \in \mathbf{Z}$.

(iv) If $p \mid B$, $p \nmid a$ then $p \nmid x$.

(v)   If $p \mid B$, $p \nmid x$ then $p \nmid a$ [*Hint.* Consider repeated duplication.]

(vi) If $3^l \parallel b$, $3^m \parallel a$ then $l = 0, 1$ or $l = 2$, $m \geq 1$. [*Hint.* If $l = 3$ deduce that either $3 \nmid x$ or $3^6 \mid B$]

(vii) Hence $w \in \mathbf{Z}_3$.

(viii) $w \in \mathbf{Z}_2$.

(ix) $w \in \mathbf{Z}_q$ for $q \mid B$, $q \neq 2, 3$.

(x)   Hence $w \in \mathbf{Z}$.

4. Show that
$$X^3 + Y^3 + dZ^3 = 0$$
is birationally equivalent to
$$Y^2 = X^3 - 2^4.3^3.d^2$$
If $d > 0$, $d \in \mathbf{Z}$ is cube free, deduce from the preceding exercise that the only cases of torsion are
$$d = 1, \quad (1, 0, -1) \text{ and } (0, 1, -1) \text{ of order 3.}$$
$$d = 1, \quad (1, 1, -1) \text{ of order 2.}$$
Compare with results of §6 on exceptional points.

5. Let $s \in \mathbf{Q}$. Show that if there is one $k \in \mathbf{Q}$ such that
$$X^3 + sX + k = 0$$
has 3 rational roots, then there are infinitely many.
[*Hint.* Let u be a rational root. Find the condition, in terms of s, u, k: that the two remaining roots are rational.]

6. Let $k \in \mathbf{Q}$, $k \neq 0$. Show that if there are two $s \in \mathbf{Q}$ such that
$$X^3 + sX + k = 0$$
has 3 rational roots, then there are infinitely many.

13

Finite Basis Theorem. Strategy and comments

The objective of the next few sections is the following.

Theorem 1. *The group $\mathfrak{G}$ of rational points on an elliptic curve defined over $\mathbb{Q}$ is finitely generated.*

The theorem is due to Mordell and it was generalized to number fields by Weil. It is usually referred to as the Mordell (or Mordell-Weil) *Finite Basis Theorem.*

For example[11], when C is

$$Y^2 = X(X^2 + 877)$$

the group $\mathfrak{G}$ is generated by $(0,0)$ of order 2 and $(u/v, r/s)$ of infinite order, where

$u = 37\ 5494\ 5281\ 2716\ 2193\ 1055\ 0406\ 9942\ 0927\ 9234\ 6201$,

$v = 6215\ 9877\ 7687\ 1505\ 4254\ 6322\ 0780\ 6972\ 3804\ 4100$,

$r = 256\ 2562\ 6798\ 8926\ 8093\ 8877\ 6834\ 0455\ 1308\ 9648\ 6691\ 5320$
$\quad 4356\ 6034\ 6478\ 6949$,

$s = 4900\ 7802\ 3219\ 7875\ 8895\ 9802\ 9339\ 9592\ 8925\ 0960\ 6161\ 6470$
$\quad 7799\ 7926\ 1000$.

The proof of Theorem 1 subdivides into two parts requiring different ideas and techniques.

---

[11] A. Bremner, J.W.S. Cassels: On the equation $Y^2 = X(X^2 + p)$. *Math. Comp.* **42** (1984), 257-264.

(i) The "weak finiteness theorem" that

$$\mathfrak{G}/2\mathfrak{G}$$

is finite. The proof depends on the construction of a map of $\mathfrak{G}/2\mathfrak{G}$ into a finite group. The proof is in some ways easier if $\mathfrak{G}$ has a point of order 2 and we do this first. For this we need to know about isogenies.

It is rather remarkable that the proof of the weak theorem is not constructive - that is, it does not give an infallible procedure, even in principle, for determining $\mathfrak{G}/2\mathfrak{G}$. Even today no algorithm is known.

(ii) The second part of the proof of the finite basis theorem is a "descent". Suppose that we have a set of representatives $\mathbf{b}_1, \ldots, \mathbf{b}_r$ of the classes of $\mathfrak{G}/2\mathfrak{G}$. Let a be any point. Then there is some s, $1 \leq s \leq r$ such that

$$\mathbf{a} - \mathbf{b}_s \in 2\mathfrak{G},$$

i.e.
$$\mathbf{a} = \mathbf{b}_s + 2\mathbf{c}, \quad \mathbf{c} \in \mathfrak{G}. \tag{*}$$

The *height* measures the size of the numbers involved in a point of $\mathfrak{G}$. For example if $\mathbf{x} = (x, y)$ and $x = u/v$ with $u, v \in \mathbf{Z}$ in its lowest terms, we can take $H(\mathbf{x}) = \max(|u|, |v|)$ (absolute values). Now it follows from (*) that $H(\mathbf{c}) < H(\mathbf{a})$; at least if $H(\mathbf{a})$ is greater than some H_0. It follows that $\mathfrak{G}$ is generated by the $\mathbf{b}_s$ and the finitely many a with $H(\mathbf{a}) \leq H_0$.

We conclude this section by giving one of Fermat's own descent arguments.

He wished to show that there are no integer solutions of

$$X^4 + Y^4 = Z^4 \quad X \neq 0, Y \neq 0.$$

This is a curve of genus 3 (not that Fermat knew about the genus), but he remarked that it is enough to disprove

$$X^4 + Y^4 = Z^2 \quad X \neq 0, Y \neq 0 \tag{*}$$

On writing (*) in the shape

$$(Z/Y^2)^2 = 1 + (X/Y)^4$$

one sees that we have an elliptic curve, though not given in canonical form. However, following Fermat, we consider integer solutions of (*).

If (*) has an integral solution, we take one (x, y, z) for which

$$\max(|x|, |y|)$$

is > 0 and as small as possible. (|| is the absolute value). Then x, y, z have no common factor, and indeed are coprime in pairs. Since $x^4 \equiv 1$ mod 4 if x is odd, one of x, y must be odd and the other even. We

suppose that
$$2 \mid x, \qquad 2 \nmid y, \qquad 2 \nmid z.$$

Write (*) in the shape
$$(z + y^2)(z - y^2) = x^4.$$

Since z, y are both odd, the two factors on the left are divisible by 2 but only one is divisible by 4. Hence (taking $z > 0$) we have two possibilities, where $u, v \in \mathbf{Z}$:

	First Case	Second Case
$z + y^2 =$	$8u^4$	$2u^4$
$z - y^2 =$	$2v^4$	$8v^4$

The first case gives
$$y^2 = 4u^4 - v^4,$$

which is impossible mod 4. Hence we have the second case:
$$y^2 = u^4 - 4u^4.$$

Now
$$(u^2 + y)(u^2 - y) = 4v^4,$$

and so
$$u^2 + y = 2r^4$$
$$u^2 - y = 2s^4$$

for some $r, s \in \mathbf{Z}$. Hence
$$r^4 + s^4 = u^2.$$

This is another solution of (*). Further,
$$x^4 = 16u^4v^4 = 16u^4r^4s^4.$$

Hence $rs \neq 0$ and
$$\max(|r|, |s|) < |x| \leq \max(|x|, |y|).$$

This contradicts the assumed minimality of the original solution, and so we have a contradiction.

Note that $(r, s, u) \to (x, y, z)$ is multiplication by 2. Thus Fermat's descent is essentially a converse of Diophantos' ascent.

Note also that multiplication by 2 has been divided into two steps via another curve
$$X^4 - 4Y^4 = Z^2.$$

This is the phenomenon of isogeny, which we explore in the next section.

§13. Exercises

1. Let C : $Y^2 = X^3 + AX + B$ be defined over $\mathbf{Q}$. Let $\mathbf{Q}(\sqrt{d})$ be

a quadratic extension of $\mathbf{Q}$ and let the non-trivial automorphism be denoted by ($'$). Let $\mathbf{x}$ be a point of C defined over $\mathbf{Q}(\sqrt{d})$. Show that $\mathbf{x} + \mathbf{x}'$ is defined over $\mathbf{Q}$ and that $\mathbf{x} - \mathbf{x}' = (u, v)$ where u and $v/\sqrt{d}$ are in $\mathbf{Q}$.

Deduce that the group of points on C defined over $\mathbf{Q}(\sqrt{d})$ may be determined once the groups over $\mathbf{Q}$ on C and $dY^2 = X^3 + AX + B$ are known.

2. This question assumes knowledge of the arithmetic of $\mathbf{Q}(\rho)$ where $\rho^3 = 1$, $\rho \neq 1$.

Fill in the details of the sketched proof of the

Theorem. *Let $d = q_1 q_2$ where $q_1 > 0$, $q_2 > 0$ are rational primes with $q_1 \equiv 2$ (9), $q_2 \equiv 5$ (9). Then the only rational point on*

$$C: \quad X_1^3 + X_2^3 + dX_3^3 = 0$$

is $(1, -1, 0)$.

Sketch proof.

(i) It is enough to prove that the only points on C defined over $\mathbf{Q}(\rho)$ are those with $X_3 = 0$

(ii) If $\mathbf{x} = (x_1, x_2, x_3)$ is defined over $\mathbf{Q}(\rho)$ and on the curve, without loss of generality x_1, x_2, x_3 are coprime in pairs in $\mathbf{Z}[\rho]$.

(iii) $(x_1 + x_2)(\rho x_1 + \rho^{-1} x_2)(\rho^{-1} x_1 + \rho x_2) = -q_1 q_2 x_3^3$. There are α_1, α_2, α_3, ξ_1, ξ_2, $\xi_3 \in \mathbf{Z}[\rho]$ such that *either*

$$x_1 + x_2 = \alpha_1 \xi_1^3, \qquad \rho x_1 + \rho^{-1} x_2 = \alpha_2 \xi_2^3,$$
$$\rho^{-1} x_1 + \rho x_2 = \alpha_3 \xi_3^3, \qquad \alpha_1 \alpha_2 \alpha_3 = d,$$

or

$$x_1 + x_2 = \lambda \alpha_1 \xi_1^3, \qquad \rho x_1 + \rho^{-1} x_2 = \lambda \alpha_2 \xi_2^3,$$
$$\rho^{-1} x_1 + \rho x_2 = \lambda \alpha_3 \xi_3^3, \qquad \alpha_1 \alpha_2 \alpha_3 = d$$

where $\lambda = \rho - \rho^{-1}$ $[= \sqrt{-3}]$.

(iv) $\alpha_1 \xi_1^3 + \alpha_2 \xi_2^3 + \alpha_3 \xi_3^3 = 0$, $\alpha_1 \alpha_2 \alpha_3 = d$,

(v) Any rational q_1-adic unit is congruent to a cube mod q, but ρ is not congruent to a cube. And similar for q_2.

(vi) After multiplying $\alpha_1, \alpha_2, \alpha_3$ all by ρ, or by ρ^2, if necessary, we may suppose that $\{\alpha_1, \alpha_2, \alpha_3\}$ is a permutation if $\{\pm 1, \pm 1, \pm q_1 \pm q_2\}$ or $\{\pm 1, \pm q_1, \pm q_2\}$.

(vii) The equation $\xi_1^3 + q_1 \xi_1^3 + q_2 \xi_2^3 = 0$ is impossible mod 9 [and indeed mod λ^3].

(viii) If $\{\alpha_1, \alpha_2, \alpha_3\}$ is a permutation of $\{\pm 1, \pm 1, \pm d\}$, then

$$|\xi_1 \xi_2 \xi_3|_\infty < |x_1 x_2 x_3|_\infty.$$

14

A 2-isogeny

An *isogeny* is a map

$$\mathcal{C} \to \mathcal{D}$$

of elliptic curves defined over the ground field and taking the specified rational point $o_{\mathcal{C}}$ on $\mathcal{C}$ into that on $\mathcal{D}$. Clearly the kernel of the isogeny, i.e. the set of points mapped into $o_{\mathcal{D}}$ is a finite group and is defined over the ground field as a whole.

In this section we consider the case when $\mathcal{C}$ has a rational point of order 2. It is convenient to modify our canonical form to

$$\mathcal{C}: \quad Y^2 = X(X^2 + aX + b),$$

the point of order 2 being $(0, 0)$. The function on the right hand side may not have a double root, so

$$b \neq 0, \qquad a^2 - 4b \neq 0.$$

We take $\mathbf{Q}$ to be the ground field. Let $\mathbf{x} = (x, y)$ be a *generic point* of $\mathcal{C}$; that is, x is transcendental and y is defined by

$$y^2 = x(x^2 + ax + b).$$

The field $\mathbf{Q}(x, y)$ is known as the *function field* of $\mathcal{C}$ over $\mathbf{Q}$.

Let

$$\mathbf{x}_1 = \mathbf{x} + (0, 0).$$

The transformation

$$\mathbf{x} \to \mathbf{x}_1$$

is an automorphism of $\mathbf{Q}(x, y)$ of order 2. We will find the fixed field.

The line through $(0,0)$ and (x,y) is

$$X = tx, \qquad Y = ty,$$

which meets $\mathcal{C}$ in $(0,0)$, $\mathbf{x}$ and $-\mathbf{x}_1 = (x_1, -y_1)$. We get

$$x_1 = b/x$$
$$y_1 = -by/x^2.$$

One invariant under $\mathbf{x} \to \mathbf{x}_1$ is clearly t^2, which is

$$t^2 = (y/x)^2 = \frac{x^2 + ax + b}{x}$$
$$= \lambda \qquad \text{(say)} \quad [= x + x_1 + a].$$

Another is

$$y + y_1 = \mu \qquad \text{(say)}.$$

To find an algebraic relation between λ, μ we compute

$$\mu^2 = y^2(1 - b/x^2)^2$$
$$= \frac{x^2 + ax + b}{x}(x^2 - 2b + b^2/x^2).$$

Here the first factor is just λ. The second is

$$(x + b/x)^2 - 4b = (\lambda - a)^2 - 4b$$
$$= \lambda^2 - 2a\lambda + (a^2 - 4b).$$

Hence

$$\mu^2 = \lambda(\lambda^2 - 2a\lambda + (a^2 - 4b)).$$

Conversely, we can express x, y in terms of λ, μ and

$$\lambda^{1/2} = y/x,$$

since

$$\lambda^{-1/2}\mu = x - b/x$$
$$\lambda = x + (b/x) + a.$$

Hence

$$x = \frac{1}{2}(\lambda + \lambda^{-1/2}\mu - a), \qquad y = \lambda^{1/2}x. \qquad (*)$$

The field extension $\mathbf{Q}(x,y)/\mathbf{Q}(\lambda,\mu)$ is of degree 2 and so by Galois theory $\mathbf{Q}(\lambda,\mu)$ is the complete field of invariants.

The point (λ, μ) is a generic point of

$$\mathcal{D}: \quad Y^2 = X(X^2 - 2aX + (a^2 - 4b)).$$

The map

$$\phi: \quad \mathcal{C} \to \mathcal{D}$$

given by

$$\mathbf{x} = (x, y) \to \lambda = (\lambda, \mu)$$

preserves the group law[12]. For let $\mathbf{a}$, $\mathbf{b}$ be points on $\mathcal{C}$ and let $f \in \mathbf{Q}(\mathbf{x})$ be a function with simple poles at $\mathbf{a}$, $\mathbf{b}$ and simple zeros at $\mathbf{o}$, $\mathbf{a+b}$. Let f_1 be the conjugate under $\mathbf{x} \to \mathbf{x}_1$. Then $f f_1 \in \mathbf{Q}(\lambda)$: as a function of λ it clearly has simple poles at $\phi(\mathbf{a})$, $\phi(\mathbf{b})$ and simple zeros at $\phi(\mathbf{o}) = \mathbf{o}$ and $\phi(\mathbf{a + b})$. Hence

$$\phi(\mathbf{a + b}) = \phi(\mathbf{a}) + \phi(\mathbf{b}).$$

The equation for $\mathcal{D}$ has the same general shape as that for $\mathcal{C}$. On repeating the process with λ and $\mathcal{D}$, we get ρ, σ with

$$\sigma^2 = \rho(\rho^2 + 4a\rho + 16b);$$

and so

$$\xi = \rho/4, \qquad \eta = \sigma/8$$

is a generic point of $\mathcal{C}$ again.

The points mapping into $(\lambda, \mu) = (0,0)$ are just the 2-division points other than $(0,0)$. Hence the kernel of the map $(x, y) \to (\xi, \eta)$ is just the 2-division points and $\mathbf{o}$. So the map must be multiplication by ± 2.

We now consider the effect of the isogeny

$$\phi: \quad \mathcal{C} \to \mathcal{D}$$

on rational points. Denote the rational points on $\mathcal{C}$, $\mathcal{D}$ by $\mathfrak{G}$, $\mathfrak{H}$ respectively.

We denote the multiplicative group of nonzero elements of $\mathbf{Q}$ by $\mathbf{Q}^*$.

Lemma 1. *Let* $(u, v) \in \mathfrak{H}$. *Then* $(u, v) \in \phi\mathfrak{G}$ *precisely when either* $u \in (\mathbf{Q}^*)^2$ *or* $u = 0$, $a^2 - 4b \in (\mathbf{Q}^*)^2$.

Proof. For $u \neq 0$, this follows by specializing $\lambda \to u$, $\mu \to v$ in (*). The point $(\lambda, \mu) = (0,0)$ comes from the points $(\alpha, 0)$ where $\alpha^2 + a\alpha + b = 0$: and $a \in \mathbf{Q}$ if and only if $a^2 - 4b \in (\mathbf{Q}^*)^2$.

This suggests the map

$$q: \quad \mathfrak{H} \to Q^*/(\mathbf{Q}^*)^2$$

given by

$$q((u, v)) = u(\mathbf{Q}^*)^2 \qquad (u \neq 0)$$
$$= (a^2 - 4b)(\mathbf{Q}^*)^2 \qquad (u = 0)$$
$$q(\mathbf{o}) = (\mathbf{Q}^*)^2.$$

---

[12] cf. §24, Lemma 1. *Exercise.* If $f = l + mx + ny$, then $ff_1 = L + M\lambda + N\mu$, where $L = l^2 - alm + bn^2$, $M = lm - bn^2$, $N = ln$.

We note that the equation
$$v^2 = u(u^2 - 2au + a^2 - 4b)$$
implies that
$$q((u,v)) = (u^2 - 2au + a^2 - 4b)(\mathbf{Q}^*)^2$$
whenever the right hand side is defined.

Lemma 2. *The map*
$$q : \mathfrak{H} \to \mathbf{Q}^*/(\mathbf{Q}^*)^2$$
is a group homomorphism.

Proof. Write the equation of $\mathcal{D}$ as
$$\mathcal{D}: \quad V^2 = U(U^2 + a_1 U + b_1).$$
Let $\mathbf{u}_j = (u_j, v_j)$ $(j = 1,2,3) \in \mathfrak{H}$ with
$$\mathbf{u}_1 + \mathbf{u}_2 + \mathbf{u}_3 = \mathbf{o},$$
so they are the intersection of $\mathcal{D}$ with a line
$$V = lU + m.$$
Substituting in the equation for $\mathcal{D}$, we have
$$U(U^2 + a_1 U + b_1) - (lU + m)^2$$
$$= (U - u_1)(U - u_2)(U_- u_3).$$
Hence
$$u_1 u_2 u_3 = m^2.$$
This implies that
$$q(\mathbf{u}_1)q(\mathbf{u}_2)q(\mathbf{u}_3) = (\mathbf{Q}^*)^2$$
except, possibly, when one of the $\mathbf{u}_j$ is $(0,0)$. The verification in this case is left to the reader.

Lemma 3. *The image of*
$$q : \quad \mathfrak{H} \to \mathbf{Q}^*/(\mathbf{Q}^*)^2$$
is finite.

Proof. Without loss of generality
$$a_1 \in \mathbf{Z}, \qquad b_1 \in \mathbf{Z}.$$
An element of $\mathbf{Q}^*/(\mathbf{Q}^*)^2$ may be written $r(\mathbf{Q}^*)^2$, where
$$r \in \mathbf{Z}, \qquad \text{square free.}$$

We show that $r(\mathbf{Q}^*)^2$ is in the image of q only when $r \mid b_1$.

Suppose that $q((u, v)) = r(\mathbf{Q}^*)^2$. Then there are $s, t \in \mathbf{Q}$ such that

$$u^2 + a_1 u + b_1 = rs^2$$

$$u = rt^2.$$

Put $t = l/m$, where

$$l, m \in \mathbf{Z}, \qquad \gcd(l, m) = 1.$$

Then, on eliminating u,

$$r^2 l^4 + a_1 r l^2 m^2 + b_1 m^4 = rn^2,$$

where $n = m^2 s \in \mathbf{Z}$.

Suppose that there is a prime p with $p \mid r$, $p \nmid b_1$. Then $p \mid m$, so $p^2 \mid rn^2$ and hence $p \mid n$ because r is square-free. Then $p^3 \mid r^2 l^4$, so $p \mid l$, contrary to $gcd(l, m) = 1$.

Putting the three lemmas together, we get the

Theorem 1. $\mathfrak{H}/\phi\mathfrak{G}$ *is finite.*

Corollary. $\mathfrak{G}/2\mathfrak{G}$ *is finite.*

Proof. Consider the exact triangle

$$
\begin{array}{ccc}
\mathcal{C} & \xrightarrow{\times 2} & \mathcal{C} \\
{}_{\phi}\searrow & & \nearrow_{\psi} \\
& \mathcal{D} &
\end{array}
$$

where $\mathfrak{H}/\phi\mathfrak{G}$ and $\mathfrak{G}/\psi\mathfrak{H}$ are both finite.

By considering in detail the equations arising in the Lemma 3, we can get more information about $\mathfrak{G}/2\mathfrak{G}$; e.g. by looking at the equations locally. There is, however, no local-global theorem and indeed even today there is no algorithm for deciding whether or not there is a solution. We shall come back to these questions in a late section. So one should not conclude from the fact that we can determine $\mathfrak{G}/2\mathfrak{G}$ in the examples that one can always do so.

We first enunciate more precisely what was proved.

Lemma 4. *The group $\mathfrak{H}/\phi\mathfrak{G}$ is isomorphic to the group of $q(\mathbf{Q}^*)^2$ in $\mathbf{Q}^*/(\mathbf{Q}^*)^2$ where*

(i)   *$q \in \mathbf{Z}$ is square-free and $q \mid b_1$*
(ii) *The equation*
$$ql^4 + a_1 l^2 m^2 + (b_1/q)m^4 = n^2$$
has a solution in $l, m, n \in \mathbf{Z}$ not all 0.

Further, the point $(0,0)$ of $\mathfrak{H}$ corresponds to $q =$ the square-free kernel of b_1.

Example 1.
$$\mathcal{C}: \quad Y^2 = X(X^2 - X + 6)$$
$$\mathcal{D}: \quad Y^2 = X(X^2 + 2X - 23)$$

For $\mathfrak{H}/\phi\mathfrak{G}$ we have $q \mid (-23)$. Since -23 corresponds to $(0,0)$, we need look at only one of $q = +23$, $q = -1$, say the latter. The equation of Lemma 4 is
$$-l^4 + 2l^2 m^2 + 23m^4 = n^2$$
i.e.
$$-(l^2 - m^2)^2 + 24m^4 = n^2,$$
which is impossible in $\mathbf{Q}_3$. Hence $\mathfrak{H}/\phi\mathfrak{G}$ is generated by $(0,0)$.

For $\mathfrak{G}/\psi\mathfrak{H}$, we have $q \mid 6$, so $q = -1$ or $q = \pm 2, \pm 3, \pm 6$. Since the form $X^2 - X + 6$ is definite, we must have $q > 0$. Hence $q = 2, 3$ or 6; and 6 belongs to $(0,0)$. Thus it is enough to look at one of $2, 3$, say 2. The equation is
$$2l^4 - l^2 m^2 + 3m^4 = n^2,$$
which is seen to have the solution $(l, m, n) = (1, 1, 2)$. This corresponds to $(x, y) = (2, 4)$.

It follows that $\mathfrak{G}/\psi\mathfrak{H}$ is generated by $(0,0)$ and $(2,4)$. To find generators for $\mathfrak{G}/2\mathfrak{G}$ we need to look at the effect of ψ on the generators of $\mathfrak{H}/\phi\mathfrak{G}$. In this case $\psi(0,0) = o$, so $\mathfrak{G}/2\mathfrak{G}$ is also generated by $(0,0)$ and $(2,4)$.

Second example. This is related to Fermat's equation
$$U^4 + V^4 = V^4.$$
Then
$$Y = V^2 W^2/U^4, \qquad X = W^2/U^2$$

satisfy

$$C: \; Y^2 = X(X^2 - 1),$$

so

$$\mathcal{D}: Y^2 = X(X^2 + 4).$$

For $\mathfrak{H}/\phi\mathfrak{G}$, we have $q \mid 4$, so $q = -1, \pm 2$. Since $X^2 + 4$ is definite, we need $q > 0$, so only $q = 2$ needs to be looked at. The relevant equation is

$$2l^4 + 2m^4 = n^2,$$

which has the solution $(l, m, n) = (1, 1, 2)$, giving $(X, Y) = (2, 4)$ as the generator of $\mathfrak{H}/\phi\mathfrak{G}$. The point $(0, 0)$ is in $\phi\mathfrak{G}$.

For $\mathfrak{G}/\psi\mathfrak{H}$, we have $q \mid (-1)$. Since -1 belongs to $(0, 0)$, there is nothing to do. Then $\mathfrak{G}/\psi\mathfrak{H}$ is generated by $(0, 0)$ and $\mathfrak{G}/2\mathfrak{G}$ is generated by $(0, 0)$ and $\psi(2, 4) = (1, 0)$.

§14. Exercises

1. Find

(i)   a set of generators for $\mathfrak{G}/2\mathfrak{G}$, where $\mathfrak{G}$ is the group of rational points and

(ii)   the 2-power torsion, for the following curves

$$Y^2 = X(X^2 + 3X + 5)$$
$$Y^2 = X(X^2 - 4X + 15)$$
$$Y^2 = X(X^2 + 4X - 6)$$
$$Y^2 = X(X^2 - X + 6)$$
$$Y^2 = X(X^2 + 2X + 9)$$
$$Y^2 = X(X^2 - 2X + 9)$$

2. Invent similar questions to 1 and solve them. [*Note.* You cannot expect to determine $\mathfrak{G}/2\mathfrak{G}$ in every case, but you can majorize its order. It might be helpful to write a Mickey Mouse program to look for points with small co-ordinates.]

3. Let $C : Y^2 = X(X^2 + aX + b)$, $\mathcal{D} : Y^2 = X(X^2 + a_1 X + b_1)$ with $a_1 = -2a$, $b_1 = a^2 - 4b$.

(i)   Show that the odd torsion groups are isomorphic

(ii)   Assuming the finite basis theorem, show that the ranks [= number of generators of infinite order] are the same

(iii)   give an example to show that the orders of the groups of 2-power torsion need not be the same. Determine what the possibilities are.

4. (i)   Construct an elliptic curve with a torsion element of order 8.

(ii)   [*Harder.*] Show that no torsion element can have order 16.

(iii)   Determine all abstract groups of 2-power order which can isomorphic to the 2-power torsion of an elliptic curve. Give elliptic curves in the possible cases and give a proof of impossibility for the others.

5. (Another kind of isogeny). Let
$$C : Y^2 = X^3 + B$$
be defined over $\mathbf{Q}$ and let $\beta^2 = B$, $\beta \in \overline{\mathbf{Q}}$.

(i)   Show that $Y = \pm\beta$ are inflexions and that $2(0, \beta) = (0, -\beta)$.

(ii)   Let $\mathbf{x} = (x, y)$ be generic and put
$$\mathbf{x}_1 = \mathbf{x} + (0, \beta), \qquad \mathbf{x}_2 = \mathbf{x} + (0, -\beta).$$
Show that
$$\xi = x + x_1 + x_2, \qquad \eta = y + y_1 + y_2$$
are functions of (x, y) defined over $\mathbf{Q}$ and that
$$D : \quad \eta^2 = \xi^3 - 27B.$$

(iii)   Show that the repetition of the above map is (essentially) multiplication by 3.

(iv)   Denote by $\mathfrak{G}$, $\mathfrak{H}$ the groups of rational points on C, D respectively. Denote by $\mathbf{Q}(\beta)^*$ the multiplicative group of non zero elements of $\mathbf{Q}(\beta)$. If $(x, y) \in \mathfrak{G}$ and
$$y + \beta \in \{\mathbf{Q}(\beta)^*\}^3$$
show that $\mathbf{x}$ is in the image of $\mathfrak{H}$ under $D \to C$.
[*Hint.* Put $y + \beta = (u + v\beta)^3$ and equate the coefficients of β.]

(v)   Show that
$$(x, y) \to (y + \beta)\{\mathbf{Q}(\beta)^*\}^3$$
is a homomorphism
$$\mu : \quad \mathfrak{G} \to \mathbf{Q}^*(\beta)/\{\mathbf{Q}(\beta)^*\}^3$$
whose kernel is the image of $\mathfrak{H}$.

(vi)   (Requires algebraic number theory). Show that the image of μ is finite [*Hint.* cf. §16].

(vii)   Deduce that $\mathfrak{G}/3\mathfrak{G}$ is finite.

15

The weak finite basis theorem

In this section we show that $\mathfrak{G}/2\mathfrak{G}$ is finite, where $\mathfrak{G}$ is the group of rational points on the elliptic curve

$$Y^2 = F(X),$$

where

$$F(X) = X^3 + AX + B, \qquad 4A^3 + 27B^2 \neq 0.$$

The argument has similarities with that in the previous section, where we made the addition assumption that $F(X)$ has a rational root.

Here we treat in a uniform manner the cases when $F(X)$ has 3 rational roots, one rational root, no rational root. We work with the commutative ring

$$\mathbf{Q}[\Theta] = \mathbf{Q}[T]/F(T),$$

where T is a variable and Θ is the image of T. Then $\mathbf{Q}[\Theta]$ is the direct sum of as many fields as $F(T)$ has irreducible factors[13].

There is a *norm* map

$$\text{Norm}: \quad \mathbf{Q}[\Theta] \to \mathbf{Q}$$

defined as follows. Let $\alpha \in \mathbf{Q}[\Theta]$. The map

$$\xi \to \alpha\xi \qquad \xi \in \mathbf{Q}[\Theta]$$

---

[13] The preceding section has proved the weak finite basis theorem when $F(T)$ has a rational root, so it would be enough to consider the case when $\mathbf{Q}[\Theta]$ is a field. This brings some minor simplifications to the proof.

takes $\mathbf{Q}[\Theta]$ into itself. If $\mathbf{Q}[\Theta]$ is regarded as a 3-dimensional vector space over $\mathbf{Q}$, the map is linear and its determinant is defined to be $\text{Norm}(\alpha)$. Clearly

$$\text{Norm}(\alpha\beta) = \text{Norm}(\alpha)\,\text{Norm}(\beta);$$

and α is *invertible* (i.e. has an inverse) precisely when $\text{Norm}(\alpha) \neq 0$. It is readily checked that

$$\text{Norm}(a - \Theta) = F(a) \qquad (a \in \mathbf{Q}).$$

Denote by $\mathbf{Q}[\Theta]^*$ the multiplicative group of invertible elements of $\mathbf{Q}[\Theta]$. We shall work with the group

$$\mathcal{M} \subset \mathbf{Q}[\Theta]^*/(\mathbf{Q}[\Theta]^*)^2$$

consists of the $\alpha(\mathbf{Q}[\Theta]^*)^2$ for which $\text{Norm}\,\alpha \in (\mathbf{Q}^*)^2$.

There is a map

$$\mu : \quad \mathfrak{G} \to \mathcal{M}$$

defined as follows.

(i) $\mu(\mathbf{o}) = 1(\mathbf{Q}[\Theta]^*)^2$

(ii) if $\mathbf{a} = (a, b) \in \mathfrak{G}$, $b \neq 0$, then

$$\mu(\mathbf{a}) = (a - \Theta)(\mathbf{Q}[\Theta]^*)^2$$

(iii) if[14] $\mathbf{a} = (a, 0)$, then $F(a) = 0$, so one of the summands in the expression of $\mathbf{Q}[\Theta]$ as a sum of fields is a copy of $\mathbf{Q}$ arising from the map $\Theta \to a$. Hence this component of $a - \Theta$ is 0. We replace (*patch*) this component with *any* element of $\mathbf{Q}^*$ such that the norm of the new element of $\mathbf{Q}[\Theta]$ is in $(\mathbf{Q}^*)^2$.

Lemma 1. *The map μ is a group homomorphism.*

Proof. Let $\mathbf{a}_j = (a_j, b_j)$ $(j = 1, 2, 3)$ be elements of $\mathfrak{G}$ with

$$\mathbf{a}_1 + \mathbf{a}_2 + \mathbf{a}_3 = 0,$$

so that they lie on a line

$$Y = lX + m \qquad l, m \in \mathbf{Q}.$$

Then

$$F(X) - (lX + m)^2 = (X - a_1)(X - a_2)(X - a_3).$$

Replace X by Θ:

$$(a_1 - \Theta)(a_2 - \Theta)(a_3 - \Theta) = (l\Theta + m)^2.$$

---

[14] cf. preceding footnote.

If all the $b_j \neq 0$, then the $a_1 - \Theta$ are invertible and we are done.

It remains to deal with the case when $F(T)$ is reducible and at least one of the roots is among the a_j. If only one of the roots, e (say), of $F(T)$ is among the a_j, then $\mathbf{Q}[\Theta]$ is a direct sum $K_1 \oplus K_2$ or $K_1 \oplus K_2 \oplus K_3$ of fields, where K_1 is the copy of $\mathbf{Q}$ given by $\Theta \to e$. The given proof shows that the Lemma holds for the components in K_j ($j \neq 1$). Since we have patched things so that the norms are always a square, the Lemma must hold for the K_1-components as well.

The remaining case is when all the b_j are 0 and the a_j are the roots of $F(T)$. Then $\mathbf{Q}[\Theta]$ is the direct sum of three copies K_j of $\mathbf{Q}$ by $\Theta \to a_j$ ($j = 1, 2, 3$). The components of $\Theta - a_1$ in K_2, K_3 are $a_2 - a_1$, $a_3 - a_1$ respectively. Hence the patch for the zero compound of $\Theta - a_1$ in K_1 is $(a_2 - a_1)(a_3 - a_1)(\mathbf{Q}^*)^2$. Now the truth of the Lemma follows by direct calculation.

Lemma 2. *The kernel of μ is $2\mathfrak{G}$.*

Proof. Since $\mathcal{M}$ has exponent 2, the kernel certainly contains $2\mathfrak{G}$. We have to show it is no bigger.

Suppose that

$$\mu(\mathbf{a}) = (\mathbf{Q}^*[\Theta])^2, \qquad \mathbf{a} = (a, b).$$

Then[15]

$$a - \Theta = (p_2 \Theta^2 + p_1 \Theta + p_0)^2$$

for some p_0, p_1, $p_2 \in \mathbf{Q}$. Further,

$$p_2 \neq 0,$$

since Θ does not satisfy any equation of degree < 3.

We can find s_0, s_1, r_0, $r_1 \in \mathbf{Q}$ such that

$$(s_1 \Theta + s_0)(p_2 \Theta^2 + p_1 \Theta + p_0) = r_1 \Theta + r_0,$$

since the vanishing of the Θ^2-component on the right hand side is a linear condition on s_0, s_1. If $s_1 = 0$, $s_0 \neq 0$, we would have $p_2 = 0$. Hence, without loss of generality,

$$s_1 = -1.$$

Now

$$(s_0 - \Theta)^2 (a - \Theta) = (r_1 \Theta + r_0)^2.$$

---

[15] A moment's consideration shows that this statement remains true when $b = 0$, though then $p_2 \Theta^2 + p_1 \Theta + p_0$ is not invertible.

On replacing Θ by an indeterminate X, we have

$$(r_1 X + r_0)^2 - (s_0 - X)^2 (a - X) = F(X),$$

since the coefficient of X^3 is 1.

Hence the line

$$Y = r_1 X + r_0$$

meets $Y^2 = F(X)$ in $(a, \pm b)$ and (s_0, t) (twice) for some t. It follows that $(a, b) \in 2\mathfrak{G}$, as required.

Theorem 1. $\mathfrak{G}/2\mathfrak{G}$ *is finite.*

Proof. It is enough to show that the image of $\mu : \mathfrak{G} \to \mathcal{M}$ is finite.

We may suppose without loss of generality that

$$A, B \in \mathbf{Z}.$$

Let $\mathbf{x} = (x, y) \in \mathfrak{G}$. Then $y^2 = F(x)$ implies that

$$x = r/t^2, \qquad y = s/t^3,$$

where

$$r, s, t \in \mathbf{Z}, \qquad \gcd(r, t) = \gcd(s, t) = 1,$$

and

$$s^2 = r^3 + Art^4 + Bt^6.$$

To illustrate ideas, suppose for now that the roots e_1, e_2, e_3 of $F(X)$ are rational, and so in $\mathbf{Z}$. Then

$$s^2 = (r - e_1 t^2)(r - e_2 t^2)(r - e_3 t^2). \qquad (*)$$

Now

$$\gcd\{(r - e_1 t^2), (r - e_2 t^2)\}$$

divides $(e_1 - e_2)t^2$ and $(e_1 - e_2)r$, so divides $(e_1 - e_2)$: and similarly for the other pairs of factors. Hence and by $(*)$

$$r - e_j = d_j v_j^2,$$

where d_j is square free,

$$d_j | (e_1 - e_2)(e_2 - e_3)(e_3 - e_1),$$

and

$$d_1 d_2 d_3 = \text{square}.$$

There are thus only finitely many sets $\{d_1, d_2, d_3\}$; which proves the theorem in this case.

Before leaving this special case, we note that (v_1, v_2, v_3, t) lies on the curve given redundantly by

$$\mathcal{D}: \begin{cases} (e_1 - e_2)t^2 = d_2 v_2^2 - d_1 v_1^2 \\ \\ (e_2 - e_3)t^2 = d_3 v_3^2 - d_2 v_2^2 \\ \\ (e_3 - e_1)t^2 = d_1 v_1^2 - d_3 v_3^2 \end{cases}$$

We may therefore get further information about $\mathfrak{G}/2\mathfrak{G}$ by looking whether there is a rational point on $\mathcal{D}$. In particular, one may be able to show that there is not a rational point by local considerations.

Now consider the general case. Denote the roots of $F(X)$ by $\varepsilon_j \in \overline{\mathbf{Q}}$ ($j = 1, 2, 3$). We work[16] in $K = \mathbf{Q}(\varepsilon_1, \varepsilon_2, \varepsilon_3)$. As in the rational case, the ideal $[r - \varepsilon_1 t^2, r - \varepsilon_2 t^2]$ divides $\varepsilon_1 - \varepsilon_2$. Hence each principal ideal $[r - \varepsilon_j t^2]$ is a square up to one of a finite number of ideal factors. The finiteness of the class-number and the finite generation of units now imply that

$$r - \varepsilon_j t^2 = \delta_j \lambda_j^2,$$

where δ_j, $\lambda_j \in \mathbf{Q}(\varepsilon_j)$ and $\{\delta_1, \delta_2, \delta_3\}$ is from a finite set. This clearly shows that the image of μ is finite and so completes the proof of the Theorem.

We now find a curve $\mathcal{D}$ with properties analogous to those of the $\mathcal{D}$ constructed above in the case when the roots are rational.

We have shown that if $(x, y) \in \mathfrak{G}$, then

$$x - \Theta = \delta \lambda^2,$$

where δ, $\lambda \in \mathbf{Q}[\Theta]$ and δ is one of a finite set. Write

$$\lambda = v_0 + v_1 \Theta + v_2 \Theta^2 \qquad (v_j \in \mathbf{Q}).$$

Then the right hand side becomes

$$H_0(\mathbf{v}) + H_1(\mathbf{v})\Theta + H_2(\mathbf{v})\Theta^2,$$

where $H_j(\mathbf{v}) \in \mathbf{Q}[\mathbf{v}]$ is a quadratic form depending on δ. Hence there is a rational point $(\mathbf{v}, t)$ on

$$\mathcal{D}: \begin{cases} H_2(\mathbf{v}) = 0 \\ \\ H_1(\mathbf{v}) = -t^2. \end{cases}$$

---

[16] This is the only place where the use of algebraic number theory is unavoidable. If she does not know the theory, the reader should take it on trust that it is very like the rational case. But see next footnote.

Again, we can get further information on $\mathfrak{G}/2\mathfrak{G}$ by examining whether there is a point on $\mathcal{D}$ everwhere locally. If not, then δ cannot occur. If there is, we can make s further useful transformation. If $\mathcal{D}$ has a point everwhere locally, there is always a point everwhere locally on the conic

$$H_2(\mathbf{v}) = 0.$$

There is a point on $H_2(\mathbf{v}) = 0$ globally by Theorem 1 of §3, and so (see exercises)

$$H_2(\mathbf{v}) = hL^2 - MN$$

identically, for some $h \in \mathbf{Q}$ and some linear forms

$$L(\mathbf{v}), \; M(\mathbf{v}), \; N(\mathbf{v}) \in \dot{\mathbf{Q}}[\mathbf{v}].$$

Hence the rational points on $H_2(\mathbf{v}) = 0$ can be parametrized in terms of r, s (say) by

$$v_j = V_j(r, s) \qquad (j = 1, 2, 3)$$

where the $V_j(r, s) \in \mathbf{Q}[r, s]$ are quadratic forms.

It follows that $\mathcal{D}$ is birationally equivalent to

$$\mathcal{D}' : \quad t^2 = G(r, s),$$

where G is a quartic form. It would be possible to describe the possible equivalence classes of quartic forms G in terms of its invariants instead of the detour through algebraic number theory[17]. In fact this is what Birch and Swinnerton-Dyer did in their historic computations. [B.J. Birch and H.P.F. Swinnerton-Dyer. Notes on elliptic curves I, II. *J. reine angew. Math.* **212** (1963), 7-25; **218** (1965), 79-108].

We conclude this section by looking at a couple of examples.

First Example.

$$Y^2 = X(X^2 - 1).$$

We considered this already as an example of isogenies. Let $(r/t^2, s/t^3)$ be on the curve, so

$$r(r + t^2)(r - t^2) = s^2.$$

The greatest common divisor of $(r \pm t^2)$ is 1 or 2: that of r and $r + t^2$ or $r - t^2$ is 1. Further,

$$r + t^2 > r > r - t^2.$$

---

[17] This line of argument proves the finiteness of $\mathfrak{G}/2\mathfrak{G}$ without algebraic number theory at the expense of a fairly substantial study of binary quartic forms.

Hence if
$$r + t^2 = d_1 v_1^2, \qquad r = d_2 v_2^2, \qquad r - t^2 = d_3 v_3^2$$
with the d_j square free, the only possibilities are
$$(d_1, d_2, d_3) = (1, 1, 1),$$
$$(2, 1, 2),$$
$$(1, -1, -1),$$
$$(2, -1, -2).$$

These are all realized by the points of order 2. Hence $\mathfrak{G}/2\mathfrak{G}$ is generated by them.

Second example. Most applications require algebraic number theory. We give one such application, to which we will want to refer later.

The curve
$$X^3 + Y^3 + 60Z^3 = 0$$
is birationally equivalent to
$$Y^2 = X^3 - 3^3(30)^2.$$
We shall work in $\mathbf{Q}(\delta)$ where $\delta^3 = 30$. This has class number $h = 3$ and fundamental unit[18] $\eta = 1 + 9\delta - 3\delta^2$.

The roots of $F(X) = X^3 - 3^3(30)^2$ are $3\delta^2$, $3\rho\delta^2$, $3\rho^2\delta^2$, where $\rho^3 = 1$. In our usual notation, if $(r/t^2, s/t^3) \in \mathfrak{G}$, a prime common ideal divisor of any two of
$$r - 3\delta^2 t^2, \qquad r - 3\rho\delta^2 t^2, \qquad r - 3\rho^2\delta^2 t^2$$
must divide 2.3.5. Since 2, 3, 5 ramify completely, $r - 3\delta^2 t^2$ must be a perfect ideal square.

In the real embedding, clearly
$$r - 3\delta^2 t^2 > 0,$$
and so either $r - 3\delta^2 t^2 = \alpha^2$ or $r - 3\delta^2 t^2 = \eta\alpha^2$ for some $\alpha \in \mathbf{Q}(\delta)$.

We disprove the second alternative. Put
$$\alpha = u + v\delta + w\delta^2.$$
Equating coefficients of powers of δ in
$$r - 3\delta^2 t^2 = \eta\alpha^2,$$

---

[18] As it can be mildly troublesome to check that a unit is fundamental, all we actually use is that $\eta > 0$ in the real embedding and η is not a square. The last fact follows by looking at η modulo $[2 - \delta, 11]$.

we get

$$0 = 9u^2 + 2uv - 9ov^2 - 180uw + 540vw + 30w^2$$
$$-3t^2 = -3u^2 + 18uv + v^2 + 2uw - 180vw + 270w^2.$$

On putting

$$u = -28e + 90f,$$
$$v = -9e + 29f,$$
$$w = g - 3e + 9f,$$

in the first equation, it becomes

$$0 = 30g^2 - 4ef.$$

Hence there are m, n such that

$$e : f : g = m^2 : 30n^2 : 2mn.$$

On substituting in the second equation, we get for some l:

$$-3l^2 = 3m^4 - 112m^3n + 1620m^2n^2$$
$$- 10800mn^3 + 27900n^4.$$

But this is impossible in $\mathbf{Q}_2$. (Consider $|n|_2 \le |m|_2$ and $|n|_2 > |m|_2$ separately).

Hence $\mathfrak{G}/2\mathfrak{G}$ is the trivial group.

§15. Exercises

1. Determine the 2-power torsion and sets of representatives $\mathfrak{G}/2\mathfrak{G}$ for $Y^2 = F(X)$ in the following cases.

(i) $F(X) = X(X - 3)(X + 4)$
(ii) $F(X) = X(X - 1)(X + 3)$
(iii) $F(X) = X(X + 1)(X - 14)$
(iv) etc.
(v)   etc.

2. (i) Give the general form of an elliptic curve with a rational point of order 4. [*Hint*: use isogenies.]

(ii)   Show that an elliptic curve cannot have two independent rational points of order 4, i.e. points **a**, **b** such that $4\mathbf{a} = 4\mathbf{b} = \mathbf{o}$, $2\mathbf{a} \ne 2\mathbf{b}$, $2\mathbf{a} \ne \mathbf{o}$, $2\mathbf{b} \ne \mathbf{o}$.

3. Make more explicit the algorithms of the text for the case of rational roots. More precisely, let

$$F(X) = (X - e_1)(X - e_2)(X - e_3)$$

where $e_j \in \mathbf{Q}$ and let

$$\sigma_j : \quad \Theta \to e_j \quad\quad (j = 1, 2, 3)$$

be the homomorphisms of $\mathbf{Q}[\Theta]$ into $\mathbf{Q}$.

(i)   For given $t_1, t_2, t_3 \in \mathbf{Q}$, find an explicit $\lambda = l_0 + l_1\Theta + l_2\Theta^2$ $(l_j \in \mathbf{Q})$ with

$$\sigma_j(\lambda) = t_j \quad\quad (j = 1, 2, 3).$$

Show that λ is unique.

(ii)   Let $x \in \mathbf{Q}$ be such that

$$x - e_j = t_j^2 \quad\quad (j = 1, 2, 3).$$

Show that the λ constructed in (i) satisfy $\lambda^2 = x - \Theta$.

(iii)   Find in terms of the t_j, e_j the $s_0 \in \mathbf{Q}$ such that

$$(s_0 - \Theta)\lambda = r_0 + r_1\Theta \quad\quad \text{(say)}$$

has no terms in Θ^2.

(iv)   Show that $(x, t_1 t_2 t_3) = 2(s_0, ?)$ for some $? \in \mathbf{Q}$.

(v)   On replacing t_j by $\pm t_j$ (independent signs) show that one gets in general further $\mathbf{x}_1 \in \mathfrak{G}$ with $2\mathbf{x}_1 = \mathbf{x}$. What is the relation between the different $\mathbf{x}_1$?

(vi)   Using the above with $F(X) = X(X - 3)(X + 5)$ and $\mathbf{x} = (4, 6)$, find all the $\mathbf{x}_1$ with $2\mathbf{x}_1 = \mathbf{x}$.

4. [Fermat, Euler]. By transforming it to canonical form, or otherwise, show that the only rational points (x_1, x_2, x_3, x_4) on the curve

$$X_1^2 - 2X_2^2 + X_3^2 = 0, \quad X_2^2 - 2X_3^2 + X_4^2 = 0$$

are those with $x_1^2 = x_2^2 = x_3^2 = x_4^2$.

If $n_1 < n_2 < n_3 < n_4$ are integers in arithmetic progression, deduce that they cannot all be perfect squares.

16

Remedial mathematics. Resultants.

As they are often not included nowadays in undergraduate courses, we give here some basic facts about resultants and discriminants. The ground field is arbitrary.

Let

$$F(X) = f_n X^n + f_{n-1} X^{n-1} + \ldots + f_0$$
$$G(X) = g_m X^m + g_{m-1} X^{m-1} + \ldots + g_0$$

be polynomials. The polynomials

$$
\left.
\begin{aligned}
&F(X) \\
&X F(X) \\
&\quad\vdots \\
&X^{m-1} F(X) \\
&G(X) \\
&\quad\vdots \\
&X^{n-1} G(X)
\end{aligned}
\right\}
\qquad (*)
$$

can be regarded as $m+n$ linear forms in the $m+n$ variables $X^{m+n-1}, \ldots, 1$ (the "forgetful functor"). The determinant $R(F, G)$ is the *resultant* of F, G. It is defined only up to sign.

By eliminating $X^{m+n-1}, \ldots, X$ determinantally, we express $R(F, G)$ as a linear combination of the rows (*), that is

$$A(X)F(X) + B(X)G(X) = R(F, G), \qquad (1)$$

where $A(X)$, $B(X)$ have degrees $\leq m - 1$, $\leq n - 1$ respectively. If

F, G have coefficients in a ring, say $\mathbf{Z}$, then $R(F, G) \in \mathbf{Z}$ and $A(X)$, $B(X) \in \mathbf{Z}[X]$.

If $F(X)$, $G(X)$ have a common zero x (in the algebraic closure), then (1) implies that $R(F, G) = 0$. Conversely, suppose that $R(F, G) = 0$. Then the (*) are linearly dependent, and so there are $A(X)$, $B(X)$ of degrees $\leq m - 1$, $n - 1$, not both zero[19], such that

$$A(X)F(X) + B(X)G(X) = 0.$$

If we suppose that $F(X)$, $G(X)$, have precise degrees n, m (i.e. $f_n \neq 0$, $g_m \neq 0$), it follows that $F(X)$, $G(X)$ have a common factor, and so a common zero in the algebraic closure.

If $f_n = g_m = 0$, then clearly $R(F, G) = 0$. If $f_n \neq 0$ but $g_m = 0$, then clearly

$$R(F, G) = f_n R(F, G^*),$$

where

$$G^* = g_{m-1}X^{m-1} + \ldots + g_0.$$

Hence the elegant formulation is that the homogeneous forms

$$f_n X^n + f_{n-1}X^{n-1}U + \ldots + f_0 U^n$$
$$g_m X^m + \ldots + g_0 U^m$$

have a common zero $(x, u) \neq (0, 0)$ in the algebraic closure if and only if $R(F, G) = 0$.

Revert to the inhomogeneous polynomials and let

$$F(X) = f_n \prod_j (X - \theta_j)$$

$$G(X) = g_m \prod_R (X - \phi_k).$$

If $f_m, g_n, \theta_1, \ldots, \theta_n, \phi_1, \ldots, \phi_m$ are taken as variables, $R(F, G)$ is a polynomial in them. It vanishes when any θ_j is equal to any ϕ_k. Hence and from considerations of degree,

$$R(F, G) = \pm f_n^m f_m^n \prod_{j,k} (\theta_j - \phi_k)$$

$$= \pm f_n^m \prod G(\theta_j)$$

$$= \pm g_m^n \prod F(\phi_k).$$

---

[19] The particular $A(X)$, $B(X)$ given by the determinantal elimination which gave (1) may, of course, both be 0.

Let $H = H(X)$ be a further polynomial. Then it readily follows that
$$R(F, GH) = \pm R(F, G)R(F, H).$$

Further, if G_1, G_2 have the same degree m and $G_1 - G_2$ is divisible by F, we have
$$R(F, G_1) = \pm R(F, G_2).$$

Finally, we put $G = F'$, the (formal) derivative. Since
$$F'(\theta_i) = f_n \prod_{j \neq 1}(\theta_i - \theta_j)$$

we have
$$R(F, F') = \pm f_n^{2n-1} \prod_{i<j}(\theta_i - \theta_j)^2$$

The function on the right side with $+$ is the *discriminant* $D(F)$. It vanishes precisely when F has a multiple root.

For example, when $F(X) = X^3 + AX + B$, we have $D = 4A^3 + 27B^2$, and (1) gives

$$(6AX^2 - 9BX + 4A^2)(3X^2 + A) - (18AX - 27B)(X^2 + AX + B)$$
$$= 4A^3 + 27B^2.$$

§16. Exercises

1. Let $F(X) \in \mathbf{Z}_p[X]$ have discriminant D and let $a \in \mathbf{Z}_p$. If
$$|F(a)|_p < |D|_p,$$
show that $|F'(a)|_p \geq |D|_p$.

17

Heights. Finite Basis Theorem.

We are now in a position to introduce the notion of height, and so to complete the proof of the Finite Basis Theorem.

Let $\mathbf{u} = (u_0, \ldots, u_n)$ be a point of projective n-dimensional space over $\mathbf{Q}$. As the co-ordinates are homogeneous, we may suppose without loss of generality that

$$u_j \in \mathbf{Z}, \qquad \gcd(u_0, \ldots, u_n) = 1. \tag{1}$$

The *height* $\mathbf{H(u)}$ of $\mathbf{u}$ is defined to be

$$\mathbf{H(u)} = \max_j |u_j|$$

with the above normalization. In this section $|| = ||_\infty$ is the absolute value.

We shall mainly but not exclusively be concerned with the projective line. We identify $x \in \mathbf{Q}$ with the point $(x, 1)$ on the line, and so write

$$\mathbf{H}(x) = \max\{|u_0|, |u_1|\}$$

if $x = u_0/u_1$ with $u_0, u_1 \in \mathbf{Z}$ as a fraction in its lowest terms.

Lemma 1.

(i) *Let $D(U_0, U_1)$, $E(U_0, U_1) \in \mathbf{Q}[U_0, U_1]$ be forms of the same degree n. Let $\mathbf{u} = (u_0, u_1)$ be a point on the rational projective line, and suppose that $D(\mathbf{u})$, $E(\mathbf{u})$ do not both vanish. Then*

$$\mathbf{H}(D(\mathbf{u}), E(\mathbf{u})) \le c\mathbf{H(u)}^n,$$

where c is independent of $\mathbf{u}$.

(ii) *Suppose, further, that the resultant of D, E is not 0. Then there is a $\gamma > 0$, independent of $\mathbf{u}$, such that*

$$\mathbf{H}(D(\mathbf{u}), E(\mathbf{u})) \geq \gamma \mathbf{H}(\mathbf{u})^n.$$

Note. The additional hypothesis in (ii) is equivalent to supposing that D, E do not have a common zero over the algebraic closure $\overline{\mathbf{Q}}$.

Proof. By homogeneity, we may suppose that

$$D(U_0, U_1), E(U_0, U_1) \in \mathbf{Z}[U_0, U_1]$$

and that $\mathbf{u} = (u_0, u_1)$ is normalized by (1). Clearly

$$|D(\mathbf{u})|, |E(\mathbf{u})| \leq c\{\max(|u_0|, |u_1|)\}^n$$

for some c. In general $D(\mathbf{u})$, $E(\mathbf{u})$ will have a common factor, but in any case this implies the conclusion of (i).

Now suppose that the hypotheses of (ii) hold and let R be the resultant. Then there are homogeneous forms $L_j(U_0, U_1)$, $M_j(U_0, U_1) \in \mathbf{Z}[U_0, U_1]$ $(j = 0, 1)$ such that

$$L_j D + M_j E = R U_j^{2n-1} \qquad (j = 0, 1). \qquad (*)$$

On substituting $\mathbf{u}$ for U we deduce that

$$\gcd\{D(\mathbf{u}), E(\mathbf{u})\} | R.$$

Further, as in the proof of (i), there is a c' such that

$$|L_j(\mathbf{u})|, |M_j(\mathbf{u})| \leq c'\{\max(|u_0|, |u_1|)\}^{n-1} \qquad (j = 0, 1).$$

On substituting in (*) (with $\mathbf{u}$ for U), we obtain the conclusion of (ii) with $\gamma = 1/2c'$.

Now let $\mathbf{u}$, $\mathbf{v}$ be two points on the projective line and let

$$\mathbf{w} = (u_0 v_0, u_0 v_1 + u_1 v_0, u_1 v_1)$$
$$= (w_0, w_1, w_2) \qquad \text{(say)}.$$

Lemma 2.

$$\frac{1}{2} \leq \frac{\mathbf{H}(\mathbf{w})}{\mathbf{H}(\mathbf{u})\mathbf{H}(\mathbf{v})} \leq 2.$$

Proof. Let $\mathbf{u}$, $\mathbf{v}$ be normalized by (1). Then the right hand inequality is immediate.

It is readily verified that w_0, w_1, w_2 have no common factor, so it will be enough to show that

$$\max(|w_0|, |w_1|, |w_2|) \geq \frac{1}{2}\{\max(|u_0|, |u_1|)\}\{\max(|v_0|, |v_1|\};$$

which is a simple exercise left to the reader.

Back to the elliptic curve

$$C: \quad Y^2 = X^3 + AX + B$$

with

$$A, B \in \mathbb{Z}, \quad 4A^3 + 27B^2 \neq 0.$$

It is convenient (and conventional) to define the height $H(\mathbf{x})$ of a rational point $\mathbf{x} = (x, y)$ on C to be the height $H(x)$ of its X-co-ordinate. In other words, if $\mathbf{x} = (x, y, z)$ in homogeneous co-ordinates, we have

$$H(\mathbf{x}) = H(x, z). \qquad (\mathbf{x} \neq \mathbf{o})$$
$$H(\mathbf{o}) = 1.$$

Lemma 3. *There are constants* c_1, $\gamma > 0$ *depending only on* C *such that*

$$\gamma_1 \leq \frac{H(2\mathbf{x})}{H(\mathbf{x})^4} \leq c_1.$$

Proof. Writing $\mathbf{x} = (x, y)$, $2\mathbf{x} = (x_2, y_2)$, we have (cf. Formulary)

$$x_2 = D(x)/E(x),$$

where

$$D(x) = (3x + A)^2 - 8x(x^3 + Ax + B)$$
$$E(x) = 4(x^3 + Ax + B)$$

Now the resultant of $3x^2 + A$ and $x^3 + Ax + B$ is $4A^3 + 27B^2 \neq 0$, and the formulae of the previous section show that the resultant R of $D(x)$, $E(x)$ is a power of 4 times $(4A^3 + 27B^2)^2$. Hence the conditions of both parts of Lemma 1 apply with $x = u_0/u_1$ and $n = 4$; and the result follows.[20]

Lemma 4. *Let* $\mathbf{x}_1$, $\mathbf{x}_2 \in \mathfrak{G}$. *Then*

$$H(\mathbf{x}_1 + \mathbf{x}_2)H(\mathbf{x}_1 - \mathbf{x}_2) \leq c_2 H(\mathbf{x}_1)^2 H(\mathbf{x}_2)^2,$$

where c_2 *depends only on* C.

Proof. Write

$$\mathbf{x}_1 + \mathbf{x}_2 = \mathbf{x}_3, \qquad \mathbf{x}_1 - \mathbf{x}_2 = \mathbf{x}_4$$

---

[20] In fact in the proof of Lemma 1(ii) in this case one may take $4A^3 + 27B^2$ instead of R since a factor $4A^3 + 27B^2$ cancels from the L_j, M_j. Compare the formula (*) at the end of §12. Detailed formulae are in Silverman's book.

and $\mathbf{x}_j = (x_j, y_j)$ as usual. Then (cf. Formulary)

$$(1, x_3 + x_4, x_3 x_4) = (W_0, W_1, W_2)$$

as elements of the projective plane, where

$$W_0 = (x_2 - x_1)^2$$
$$W_1 = 2(x_1 x_2 + A)(x_1 + x_2) + 4B$$
$$W_2 = x_1^2 x_2^2 - 2A x_1 x_2 - 4B(x_1 + x_2) + A^2$$

On writing x_1, x_2 as quotients of integers and homogenizing, it is readily seen that

$$\mathbf{H}(W_0, W_1, W_2) \leq c_3 \mathbf{H}(x_1)^2 \mathbf{H}(x_2)^2$$

for some c_3. On the other hand,

$$\mathbf{H}(W_0, W_1, W_2) = \mathbf{H}(1, x_3 + x_4, x_3 x_4)$$
$$\geq \frac{1}{2} \mathbf{H}(x_3) \mathbf{H}(x_4)$$

by Lemma 2. The truth of the lemma follows with $c_2 = 2c_3$.

Corollary.

$$\text{Min}(H(\mathbf{x}_1 + \mathbf{x}_2), H(\mathbf{x}_1 - \mathbf{x}_2)) \leq c_4 H(\mathbf{x}_1) H(\mathbf{x}_2)$$

with $c_4 = c_2^{1/2}$.

In another direction we have

Lemma 5. *Let λ be given. There are only finitely many $\mathbf{x} \in \mathfrak{G}$ with $H(\mathbf{x}) \leq \lambda$.*

Proof. For $\mathbf{x} = (x, y)$ with $\mathbf{H}(x) \leq \lambda$; that is $x = u_0/u_1$ with $u_0, u_1 \in \mathbf{Z}$ and $|u_0|, |u_1| \leq \lambda$.

We are now in a position to prove the

Finite Basis Theorem. *The group $\mathfrak{G}$ of rational points is finitely generated.*

Proof. By the "weak" theorem [§15, Theorem 1], $\mathfrak{G}/2\mathfrak{G}$ is finite. Let $\mathbf{b}_1, \ldots, \mathbf{b}_s \in \mathfrak{G}$ be representatives of the classes of $\mathfrak{G}$ modulo $2\mathfrak{G}$.

Now let $\mathbf{a} \in \mathfrak{G}$. There is some j such that $\mathbf{a} \pm \mathbf{b}_j \in 2\mathfrak{G}$ for both choices of sign. By Lemma 4, Corollary, there is one choice of sign such that

$$H(\mathbf{a} \pm \mathbf{b}_j) \leq c_4 H(\mathbf{a}) H(\mathbf{b}_j).$$

Now $\mathbf{a} \pm \mathbf{b}_j = 2\mathbf{c}$, $\mathbf{c} \in \mathfrak{G}$, and so

$$H(\mathbf{a} \pm \mathbf{b}_j) \geq \gamma_1 H(\mathbf{c})^4$$

by Lemma 3. Putting everything together, we have

$$H(\mathbf{c})^4 \leq \gamma_1^{-1} c_4 H(\mathbf{a}) H(\mathbf{b}_j)$$
$$\leq \kappa H(\mathbf{a}),$$

where

$$\kappa = \gamma_1^{-1} c_4 \max_j H(\mathbf{b}_j).$$

Hence either

$$H(\mathbf{c}) \leq \frac{1}{2} H(\mathbf{a})$$

or

$$H(\mathbf{a}) \leq (16\kappa)^{1/3}$$
$$= \lambda \qquad \text{(say)}.$$

It follows readily that $\mathfrak{G}$ is generated by the $\mathbf{b}_j$ and the $\mathbf{a}$ with $H(\mathbf{a}) \leq \lambda$. But the latter are finite in number by Lemma 5.

We conclude this section with a brief review of the properties of heights.

The inequality in Lemma 4 is supplemented by one in the other direction:

$$H(\mathbf{x}_1 + \mathbf{x}_2) H(\mathbf{x}_1 - \mathbf{x}_2) \geq \gamma_2 H(\mathbf{x}_1)^2 H(\mathbf{x}_2)^2, \qquad (*)$$

where $\gamma_2 > 0$. Indeed the W_0, W_1, W_2 of the proof of Lemma 4, considered as functions of indeterminates x_1, x_2, have no common zero in the algebraic closure: for $W_0 = 0$ implies $x_2 = x_1$ and then W_1, W_2 become the functions D, E used in the proof of Lemma 3. Now (*) follows from an appropriate generalization of Lemma 1. Note that Lemma 3 is now just the case $\mathbf{x}_2 = \mathbf{x}_1$ of the extended Lemma 4.

We now move over to the *logarithmic height*

$$h(\mathbf{x}) = \log H(\mathbf{x}),$$

so that the extended Lemma 4 gives

$$|h(\mathbf{x}_1 + \mathbf{x}_2) + h(\mathbf{x}_1 - \mathbf{x}_2) - 2h(\mathbf{x}_1) - 2h(\mathbf{x}_2)| \leq c$$

for some constant c. In particular,

$$|h(2\mathbf{x}) - 4h(\mathbf{x})| \leq c.$$

It follows that

$$\hat{h}(\mathbf{x}) = \lim_{n \to \infty} h(2^n \mathbf{x})/4^n$$

exists, and satisfies

$$\hat{h}(\mathbf{x}_1 + \mathbf{x}_2) + \hat{h}(\mathbf{x}_1 - \mathbf{x}_2) = 2\hat{h}(\mathbf{x}_1) + 2\hat{h}(\mathbf{x}_2).$$

It is now an undergraduate exercise (cf. Exercises) to deduce that

$$\hat{h}(\mathbf{x}_1 + \mathbf{x}_2) - \hat{h}(\mathbf{x}_1) - \hat{h}(\mathbf{x}_2)$$

is bilinear in $\mathbf{x}_1$, $\mathbf{x}_2$; and so that $\hat{h}(\mathbf{x})$ is a quadratic form on $\mathfrak{G}$.

The function $\hat{h}(\mathbf{x})$ is called the[22] *canonical height*.

In particular,

$$\hat{h}(n\mathbf{x}) = n^2 \hat{h}(\mathbf{x}),$$

so $\hat{h}(\mathbf{x}) = 0$ if $\mathbf{x}$ is torsion: the converse holds by Lemma 5 and since $h(\mathbf{x}) - \hat{h}(\mathbf{x})$ is bounded.

§17. Exercises

1. (i) Let $a \in \mathbf{Q}$, $a \neq 0$. Show that $|a|_p = 1$ except for at most finitely many primes p and that

$$\prod_{p \text{ inc } \infty} |a|_p = 1.$$

(ii) Let $u_0, \ldots, u_n \in \mathbf{Q}$, not all 0. Show that $\max |u_j|_p = 1$ except for at most finitely many p and that

$$\prod_{p \text{ inc } \infty} \max_j |u_j|_p = \mathbf{H}(\mathbf{u})$$

is the height of the point $\mathbf{u} = (u_0, \ldots, u_n)$ in projective space.

2. (Required in text.) Let $f(x)$ be a function defined for x in a group $\mathfrak{M}$ and taking values in a field of characteristic $\neq 2$. Suppose that

$$f(x + y) + f(x - y) = 2f(x) + 2f(y)$$

for all x, $y \in \mathfrak{M}$. Show that

$$f(x) = B(x, x),$$

where $B(x, y)$ is a symmetric bilinear form.

[*Hint.* Take

$$B(x, y) = \frac{1}{2}\{f(x + y) - f(x) - f(y)\}.$$

One has to show that

$$B(x + z, y) = B(x, y) + B(z, y),$$

i.e. that

$$f(x + y + z) + f(x) + f(y) + f(z)$$
$$= f(y + z) + f(z + x) + f(x + y).$$

---

[22] There are different definitions of the canonical height. They differ by a constant factor.

One opening gambit is to observe that

$$(x + y + z) + x = (x + z) + (y + z).]$$

3. Let $C : Y^2 = X^3 + AX + B$ and suppose that $\mathbf{x_1}$, $\mathbf{x_2}$ are independent generic points. Let $\mathbf{x_3} = \mathbf{x_1} + \mathbf{x_2}$, $\mathbf{x_4} = \mathbf{x_1} - \mathbf{x_2}$. Show that

$$(x_1 - x_2)^2(x_1 + x_2 + x_3) = (y_1 - y_2)^2$$
$$(x_1 - x_2)^2(x_1 + x_2 + x_4) = (y_1 + y_2)^2.$$

Deduce that $x_1 + x_2 + x_3$, $x_1 + x_2 + x_4$ are roots of an equation

$$(x_1 - x_2)^2 T^2 + uT + v = 0,$$

where u, v are *polynomials* in x_1, x_2.

Deduce that a similar result holds for x_3, x_4.

4. (Required in text.) Let $G(X) \in \mathbb{Q}[X]$ be a nonsingular quadratic form in $\mathbf{X} = (X, Y, Z)$ and suppose that there is an $\mathbf{x} = (x, y, z) \neq (0, 0, 0)$ such that $G(\mathbf{x}) = 0$. Show that there are linear forms $L(\mathbf{X})$, $M(\mathbf{X})$, $N(\mathbf{X}) \in \mathbb{Q}[X]$ and a $d \in \mathbb{Q}^*$ such that

$$G(\mathbf{X}) = L(\mathbf{X})M(\mathbf{X}) + dN(\mathbf{X})^2.$$

[*Hints.*

(i)   Without loss of generality $\mathbf{x} = (1, 0, 0)$.

(ii) After a linear transformation on Y, Z, we may suppose $G(X) = XY + $ form in Y, Z.

(iii) Complete the square with respect to Z.]

5. Let $\hat{h}$ be the canonical height on some curve C and suppose that there are representatives of all classes of $\mathfrak{G}/2\mathfrak{G}$ in $\hat{h}(\mathbf{x}) \leq t$ for some t.

Show that $\mathfrak{G}$ is generated by the $\mathbf{a} \in \mathfrak{G}$ with $\hat{h}(\mathbf{a}) \leq t$.

Local-global for genus 1

Our attention now moves from elliptic curves to curves of genus 1 in general. In this section we give a couple of examples to show that there is no local-global principle for rational points on curves of genus 1. Subsequently, we shall give a structure to the "obstruction" to a local-global principle, namely the Tate-Shafarevich group.

The two examples we shall discuss are

$$3X^3 + 4Y^3 + 5Z^3 = 0, \tag{1}$$

due to Selmer, and

$$X^4 - 17 = 2Y^2, \tag{2}$$

due (independently) to Lind and Reichardt. The techniques we have developed so far enable us to disprove the existence of rational points. We have not, however, developed techniques to show that there are solutions everwhere locally. This is because we have left a fairly highbrow discussion of curves of genus 1 over finite fields until the end (§25). The reader may, of course, verify for any given p that there is a point defined over $\mathbf{Q}_p$ but this can never disprove the existence of some $P > 10^{10}$ (say) such that (1) or (2) has no solution in $\mathbf{Q}_P$. We shall assume without present proof that a curve of genus 1 over a finite field $\mathbf{F}_p$ always has a point defined over $\mathbf{F}_p$ (§25, Theorem 2). If, therefore, a curve such as (1) or (2) reduced mod p is still of genus 1, then there is a point mod p which can, by Lemma 1 of §10, be lifted to a point defined over $\mathbf{Q}_p$.

Assuming this[22], the only $\mathbf{Q}_p$ to be considered for (1) are $p = 2, 3, 5$ and the only ones for (2) are $p = 2, 17$. It may confidently be left to the reader to confirm that there are points for these p.

   The disproof of rational points on (1) uses

Lemma 1. *Let* a, b, c *be distinct integers* > 1 *and suppose that* $d = abc$ *is cube free. Suppose that there are* u, v, $w \in \mathbf{Z}$ *not all 0 such that*
$$au^3 + bv^3 + cw^3 = 0.$$
Then there are x, y, $z \in \mathbf{Z}$ *with* $z \neq 0$ *such that*
$$x^3 + y^3 + dz^3 = 0.$$

Proof. Let $\rho^3 = 1$, $\rho \neq 1$ and put
$$\xi = au^3 + \rho bv^3 + \rho^2 cw^3$$
$$\eta = au^3 + \rho^2 bv^3 + \rho cw^3.$$
Then
$$\xi + \eta = 3au^3$$
$$\rho\xi + \rho^2\eta = 3cw^3$$
$$\rho^2\xi + \rho\eta = 3bv^3$$
and so
$$\xi^3 + \eta^3 + d\zeta^3 = 0, \qquad \zeta = -3uvw.$$
Now the two points $(\xi, \rho\eta, \zeta)$, $(\eta, \rho^2\xi, \zeta)$ are conjugate over $\mathbf{Q}$. Hence the line joining them meets $X^3 + Y^3 + dZ^3 = 0$ in a point defined over $\mathbf{Q}$ and distinct from $(1, -1, 0)$.

Lemma 2. *The only point defined over* $\mathbf{Q}$ *on*
$$X^3 + Y^3 + 60Z^3 = 0$$
is $(1, -1, 0)$.

Proof. There is no torsion, e.g. by the discussion of exceptional points on cubic curves (§6, Lemma 1). The curve is birationally equivalent over $\mathbf{Q}$ to
$$Y^2 = X^3 - 2^4.3^3.60^2,$$

---

[22] Often, including for (1), (2), this can be proved elementarily. See e.g. A. Weil, Numbers of solutions of equations in finite fields. *Bull. Amer. Math. Soc.* **55** (1949), 497–508 (= *Collected Papers* I, 395–410.)

for which $\mathfrak{G}/2\mathfrak{G}$ is trivial by the proof at the end of the section on the weak theorem (§15, Second example). It follows from the Finite Basis Theorem that $\mathfrak{G}$ is trivial.

Theorem 1. *There are no rational points on* (1).

Proof. The last two lemmas.

The preceding proof used the theory of algebraic numbers. The next proof works entirely in the rationals.

Theorem 2. *There are no rational points on* (2).

Proof. If not, suppose (x, y) is on (2). Let $x = a/c$ as a fraction in its lowest terms. Then

$$a^4 - 17c^4 = 2b^2, \qquad \gcd(a, c) = \gcd(b, c) = \gcd(a, b) = 1.$$

Putting

$$A = a^2, \qquad C = c^2$$

we have

$$A^2 - 17C^2 = 2b^2.$$

This equation is soluble everwhere locally, so globally, and in fact

$$5^2 - 17.1^2 = 2.2^2.$$

Now

$$(5A + 17C + 4b)(5A + 17C - 4b) = 17(A + 5C)^2.$$

If there is a common odd prime divisor of the two factors on the left hand side, it divides $5A + 17C$ and $A + 5C$, so divides $8A$ and $8C$: a contradiction. The two factors on the left hand side have the same sign, which for $A = a^2$, $C = c^2$ must be positive. Hence for integers u, v there is one of two possibilities

	First Case	Second Case
$5a^2 + 17c^2 \pm 4b =$	$17u^2$	$34u^2$
$5a^2 + 17c^2 \mp 4b =$	v^2	$2v^2$
$a^2 + 5c^2 \quad =$	uv	$2uv$

In the first case

$$10a^2 + 34c^2 = 17u^2 + v^2$$

$$a^2 + 5c^2 = uv.$$

We show that this is impossible in $\mathbf{Q}_{17}$. Write $\| \| = \| \|_{17}$. By homogeneity

$$\max(|a|, |c|, |u|, |v|) = 1.$$

Since 10 is a quadratic non residue mod 17, we have

$$|a| < 1, \qquad |v| < 1.$$

The second equation gives

$$|c| < 1.$$

Finally, the first equation gives

$$|u| < 1.$$

Contradiction.

The second case gives

$$5a^2 + 17c^2 = 17u^2 + v^2$$
$$a^2 + 5c^2 = 2uv.$$

The proof that this is impossible in $\mathbf{Q}_{17}$ is similar.

§18. Exercises

1. [Uses algebraic number theory.] Supply the details of the following alternative proof of Theorem 2.

(i)   The field $\mathbf{Q}(\sqrt{17})$ has class number 1. A basis of integers is 1, $\frac{1}{2}(1 + \sqrt{17})$. A fundamental unit is $4 + \sqrt{17}$ of norm -1. The prime 2 splits into $(5 \pm \sqrt{17})/2$.

(ii) Suppose $a^4 - 17c^4 = 2b^2$ with $a, b, c \in \mathbf{Z}$, $\gcd(a, c) = 1$. Then a, c are odd and
$$\frac{a^2 \pm c^2\sqrt{17}}{2}$$
are coprime.

(iii)
$$\frac{a^2 + \sqrt{17}c^2}{2} = \left(\frac{5 \pm \sqrt{17}}{2}\right)\eta\mu^2 \qquad (*)$$

for some unit η and some integer μ.

(iv) $\eta > 0$ in both real embeddings. Hence η is a square and so can be absorbed in μ^2.

(v)   Put $\eta = 1$, $\mu = (u + v\sqrt{17})/2$ in and equate terms independent of $\sqrt{17}$. Then $4a^2 = 5(u^2 + 17v^2) \pm 34uv$, which is impossible in $\mathbf{Q}_3$ (and in $\mathbf{Q}_{17}$).

19

Elements of Galois cohomology

In the next section we have occasion to consider two curves which are both defined over $\mathbf{Q}$ and which are birationally equivalent over $\overline{\mathbf{Q}}$. Here we consider a simpler case and then set up some general machinery.

The conic

$$\mathcal{A}: \quad X_1^2 + X_2^2 = 3$$

has no rational point and so is not equivalent over $\mathbf{Q}$ to the line (co-ordinate Y, no equation). They are, however, equivalent over $\mathbf{Q}(\sqrt{3})$, for example by the equations

$$y = (x_1 - \sqrt{3})/x_2$$

$$x_1 = \frac{\sqrt{3}(1 - y^2)}{y^2 + 1}, \qquad x_2 = \frac{-2\sqrt{3}y}{y^2 + 1}.$$

Let y be transcendental, so (x_1, x_2) is a generic point of $\mathcal{A}$.

The Galois group $\mathrm{Gal}(\mathbf{Q}(\sqrt{3})/\mathbf{Q})$ can be made to act in two different ways on $\mathbf{Q}(\sqrt{3}, y) = \mathbf{Q}(\sqrt{3}, x_1, x_2)$. We can either make it act trivially on y or we can make it act trivially on (x_1, x_2).

In the first case, the non-trivial element of the Galois group induces the automorphism

$$x_1 \to -x_1, \qquad x_2 \to -x_2$$

of $\mathcal{A}$. In the second case, it induces the automorphism

$$y \to -1/y$$

of the line.

In the example, we have used the birational equivalence to identify the two function fields. In the general theory it is better to make it explicit. Let $\mathcal{A}$, $\mathcal{B}$ be two curves defined over $\mathbf{Q}$ and let

$$\phi: \quad \mathcal{A} \to \mathcal{B}$$

be a birational equivalence defined over $\overline{\mathbf{Q}}$. Let $\sigma \in \mathrm{Gal}(\overline{\mathbf{Q}}/\mathbf{Q})$. We can let σ act on the coefficients in ϕ and so obtain another birational equivalence

$$\sigma\phi: \quad \mathcal{A} \to \mathcal{B}.$$

Then

$$\theta_\sigma(\text{say}) = (\sigma\phi)\phi^{-1}: \quad \mathcal{B} \to \mathcal{B}$$

is a birational automorphism defined over $\overline{\mathbf{Q}}$.

We can act on θ_σ by $\tau \in \mathrm{Gal}(\overline{\mathbf{Q}}/\mathbf{Q})$. Then

$$\begin{aligned}
\tau\theta_\sigma &= (\tau\sigma\phi)(\tau\phi)^{-1} \\
&= [(\tau\sigma\phi)\phi^{-1}][\phi(\tau\phi)^{-1}] \\
&= \theta_{\tau\sigma}\theta_\tau^{-1}.
\end{aligned}$$

Hence

$$\theta_{\tau\sigma} = (\tau\theta_\sigma)\theta_\tau.$$

This is the[23] *cocycle identity* and $\{\theta_\sigma\}$ is a *cocycle*.

Let there be another birational equivalence

$$\phi': \quad \mathcal{A} \to \mathcal{B}$$

defined over $\overline{\mathbf{Q}}$, so

$$\phi' = \omega\phi$$

for some automorphism

$$\omega: \quad \mathcal{B} \to \mathcal{B}.$$

Then

$$\begin{aligned}
\theta'_\sigma(\text{say}) &= (\sigma\phi')(\phi'^{-1}) \\
&= \sigma\omega\theta_\sigma\omega^{-1}.
\end{aligned}$$

The two cocyles $\{\theta_\sigma\}$ and $\{\theta'_\sigma\}$ are said to be *cobounding*.

If ϕ is defined over $\mathbf{Q}$, we have

$$\theta'_\sigma = (\sigma\omega)\omega^{-1},$$

a *coboundary*. In this case $\mathcal{A}$, $\mathcal{B}$ are birationally equivalent over $\mathbf{Q}$; but we have chosen to use a different equivalence.

---

[23] We owe the rococo terminology to the topologists.

Given $\mathcal{B}$ and the cocycle $\{\theta_\sigma\}$, we can reconstruct $\mathcal{A}$ (up to a birational equivalence defined over $\mathbf{Q}$). For let $\mathbf{x}$ be a generic point of $\mathcal{B}$. We define an action $\tilde{\sigma}$ of the $\sigma \in \mathrm{Gal}(\overline{\mathbf{Q}}/\mathbf{Q})$ on $\overline{\mathbf{Q}}(\mathbf{x})$ as follows:

$$\left. \begin{array}{c} \tilde{\sigma} \text{ acts on } \overline{\mathbf{Q}} \text{ by } \sigma \\ \tilde{\sigma}\mathbf{x} = \theta_\sigma \mathbf{x}. \end{array} \right\}$$

Then for $\tau \in \mathrm{Gal}(\overline{\mathbf{Q}}/\mathbf{Q})$ we have

$$\tilde{\tau}(\tilde{\sigma}\mathbf{x}) = (\tau\theta_\sigma)(\tilde{\tau}\mathbf{x})$$
$$= (\tau\theta_\sigma)\theta_\tau \mathbf{x}$$
$$= \theta_{\tau\sigma}\mathbf{x}.$$

Thus

$$(\widetilde{\tau\sigma}) = \tilde{\tau}\tilde{\sigma}.$$

The fixed field of the $\tilde{\sigma}$ is a function field over $\mathbf{Q}$, and so gives $\mathcal{A}$ up to birational equivalence over $\mathbf{Q}$.

§19. Exercises

1. Let σ run through $\mathrm{Gal}(\overline{\mathbf{Q}}/\mathbf{Q})$. Find a cocyle $\{\theta_\sigma\}$ of birational automorphisms which twist the line into

$$X_1^2 + X_2^2 = n,$$

where n is any given element of $\mathbf{Q}^*$.

When $n = 5$ give an explicit representation of your cocycle as a coboundary.

Is your cocycle a coboundary when $n = 3$?

20

Construction of the jacobian

Let $\mathcal{D}$ be a curve of genus 1 defined over $\mathbf{Q}$. In this section we construct an elliptic curve $\mathcal{C}$, also defined over $\mathbf{Q}$, which is closely related to it. This relationship will be exploited in subsequent sections.

We must initially consider birational equivalences between elliptic curves. We work at first over a general field.

Let

$$\mathcal{C}_j: \quad Y^2 = X^3 + A_j X + B_j \qquad (j = 1, 2)$$

and let

$$\phi: \quad \mathcal{C}_1 \to \mathcal{C}_2$$

be a birational correspondence. By considering $\phi(\mathbf{x}) - \phi(\mathbf{o}_1)$ instead of $\phi(\mathbf{x})$, we may suppose without loss of generality that

$$\phi(\mathbf{o}_1) = \mathbf{o}_2,$$

where $\mathbf{o}_j$ is the point at infinity on $\mathcal{C}_j$.

The correspondence must take functions with poles of order 1 into such functions. Hence

$$\phi(X) = aX + b$$

for some a, b. Similarly

$$\phi(Y) = cY = cY + dX + e.$$

The form of the equations for $\mathcal{C}_j$ imply that

$$d = e = 0, \qquad b = 0, \qquad a^3 = c^2$$

and so

$$a = s^2, \qquad c = s^3$$

for some s. Hence

$$A_2 = s^4 A_1, \qquad B_2 = s^6 B_1. \tag{1}$$

In particular, $A_1^3/B_1^2 = A_2^3/B_2^2$ is invariant under birational equivalence.

It is conventional to work with the birational invariant

$$j = j(\mathcal{C}) = \frac{1728(4A^3)}{4A^3 + 27B^2}$$

of

$$\mathcal{C}: \quad Y^2 = X^3 + AX + B.$$

The notation j is standard. The constant $1728 = 12^3$ is suggested by the complex variable theory. Note that every elliptic curve gives a finite value of j: it is the degenerate curves that send j to infinity.

Lemma 1. *Two elliptic curves in canonical form which are birationally equivalent are related by (1) for some s. In particular, they have the same j-invariant. Further, s is in any field over which the curves and the equivalence are defined.*

Corollary. *Any birational equivalence of the elliptic curve*

$$\mathcal{C}: \quad Y^2 = X^3 + AX + B$$

taking o into o is of the form

$$Y \to s^3 Y, \qquad X \to s^2 X.$$

If $AB \neq 0$, then $s^2 = 1$. If $B = 0$, then $s^4 = 1$ and if $A = 0$, then $s^6 = 1$.

Proof. Clear from (1) with $\mathcal{C} = \mathcal{C}_1 = \mathcal{C}_2$.

Let us return to the main topic of the section. Let $\mathcal{D}$ be a curve of genus 1 defined over $\mathbf{Q}$. In general it will not have a rational point and, if it has, we may not be able to find one: but there is no difficulty in finding a point defined over $\overline{\mathbf{Q}}$. Hence there is a birational correspondence

$$\phi: \quad \mathcal{D} \to \mathcal{C}$$

defined over $\overline{\mathbf{Q}}$, where $\mathcal{C}$ is in canonical form but defined over $\overline{\mathbf{Q}}$.

Let $\sigma \in \mathrm{Gal}(\overline{\mathbf{Q}}/\mathbf{Q})$. We can act on the birational correspondence with σ and obtain

$$\sigma\phi: \quad \mathcal{D} \to \sigma\mathcal{C},$$

where

$$\sigma\mathcal{C}: \quad Y^2 = X^3 + \sigma A X + \sigma B.$$

Now C and σC are birationally equivalent over $\overline{\mathbf{Q}}$ by $(\sigma\phi)\phi^{-1}$. Hence

$$\sigma j(C) = j(\sigma C) = j(C);$$

that is, $j(C) \in \mathbf{Q}$ or equivalently $A^3/B^2 \in \mathbf{Q}$ if $AB \neq 0$. Hence by a transformation $X \to t^2 X$, $Y \to t^3 Y$ ($t \in \overline{\mathbf{Q}}$) we may suppose without loss of generality that C is defined over $\mathbf{Q}$. Of course in general ϕ is defined only over $\overline{\mathbf{Q}}$. Now

$$\theta_\sigma = (\sigma\phi)\phi^{-1}$$

is an automorphism of C.

Suppose, first, that

$$AB \neq 0.$$

Then by Lemma 1, Corollary, the automorphism θ_σ of C must be

$$\theta_\sigma : \quad \mathbf{x} \to \varepsilon_\sigma \mathbf{x} + \mathbf{a}_\sigma$$

for some point $\mathbf{a}_\sigma$ defined over $\mathbf{Q}$ and $\varepsilon_\sigma = \pm 1$. We are in the position discussed in the previous section, so

$$\theta_{\tau\sigma} = (\tau\theta_\sigma)\theta_\tau.$$

In particular, since $\varepsilon_\sigma \in \mathbf{Q}$, we have

$$\varepsilon_{\tau\sigma} = \varepsilon_\sigma \varepsilon_\tau;$$

so ε_σ is a group character.

We would like to ensure that ε_σ is always 1. If not, there is some $d \in \mathbf{Q}$ such that

$$\sigma(\sqrt{d}) = \varepsilon_\sigma \sqrt{d}.$$

The transformation

$$X \to dX, \qquad Y \to d\sqrt{d}\,Y$$

gives a new C defined over $\mathbf{Q}$: and with this we do indeed have $\varepsilon_\sigma = 1$ always.

If $AB = 0$, the same conclusion holds but the argument is a little deeper[24]. Suppose that $B = 0$, so $\varepsilon_\sigma^4 = 1$, where we define

$$\mathbf{x} \to \varepsilon\mathbf{x}$$

by

$$X \to \varepsilon^2 X, \qquad Y \to \varepsilon Y.$$

Now $\mathrm{Gal}(\overline{\mathbf{Q}}/\mathbf{Q})$ acts on ε, and

$$\varepsilon_{\tau\sigma} = (\tau\varepsilon_\sigma)\varepsilon_\tau.$$

---

[24] And may be omitted at first reading.

By "Hilbert 90" (see Exercises) there is a $\delta \in \overline{\mathbf{Q}}$ with $\delta^4 \in \mathbf{Q}$ such that
$$\sigma\delta = \varepsilon_\sigma \delta.$$
We can now modify $\mathcal{C}$, as before, so that $\varepsilon_\sigma = 1$ identically on the new $\mathcal{C}$. Similarly for $A = 0$.

Thus in every case we have found a $\mathcal{C}$ defined over $\mathbf{Q}$ and a birational equivalence
$$\phi : \quad \mathcal{D} \to \mathcal{C}$$
defined over $\overline{\mathbf{Q}}$ such that
$$(\sigma\phi)\phi^{-1} = \theta_\sigma : \quad \mathbf{x} \to \mathbf{x} + \mathbf{a}_\sigma$$
for all $\sigma \in \mathrm{Gal}(\overline{\mathbf{Q}}/\mathbf{Q})$.

To sum up, we have proved:

Theorem 1. *Let $\mathcal{D}$ be a curve of genus 1 defined over $\mathbf{Q}$. There is an elliptic curve $\mathcal{C}$ defined over $\mathbf{Q}$ and a birational equivalence*
$$\phi : \quad\quad \mathcal{D} \to \mathcal{C}$$
defined over $\overline{\mathbf{Q}}$ such that. for every $\sigma \in \mathrm{Gal}(\overline{\mathbf{Q}}/\mathbf{Q})$, the map
$$\theta_\sigma = (\sigma\phi)\phi^{-1} : \quad\quad \mathcal{C} \to \mathcal{C}$$
is of the form
$$\theta_\sigma : \quad\quad \mathbf{x} \to \mathbf{x} + \mathbf{a}_\sigma$$
for some $\mathbf{a}_\sigma \in \overline{\mathcal{O}}$.

Further, $\mathcal{C}$ is unique up to birational equivalence over $\mathbf{Q}$.

The elliptic curve $\mathcal{C}$ is the *jacobian* of $\mathcal{D}$.

Before exploring this situation further, we require some new machinery, introduced in the next section.

§20. Exercises

1. Construct the jacobian of
   (i) $Y^2 = aX^4 + bX^2 + c$ $(a, c \in \mathbf{Q}^*, b \in \mathbf{Q}, b^2 - 4ac \neq 0)$.
   (ii) $aX^3 + bY^3 + cZ^3 = 0$ $(a, b, c \in \mathbf{Q}^*)$.
   (iii) $aX^3 + bY^3 + cZ^3 + mXYZ = 0$ $(a, b, c, m \in \mathbf{Q}^*)$.
   (iv) $Y^2 = aX^4 + bX^3 + cX^2 + dX + e$.

2. Let $\mathcal{D}$ be the curve of genus 1 given by the redundant equations
   $$(e_2 - e_1)t^2 = d_1 v_1^2 - d_2 v_2^2$$
   $$(e_3 - e_2)t^2 = d_2 v_2^2 - d_3 v_3^2$$
   $$(e_1 - e_3)t^2 = d_3 v_3^2 - d_1 v_1^2,$$

where e_1, e_2, e_3 are distinct and $d_j \in \mathbf{Q}^*$, $d_1 d_2 d_3 = 1$. Show that there is a point of $\mathcal{D}$ defined over $\kappa = \mathbf{Q}(d_1^{1/2}, d_2^{1/2})$ and hence find a map

$$\phi: \quad \mathcal{D} \to \mathcal{C}$$

defined over κ into

$$\mathcal{C}: \qquad Y^2 = (X - e_1)(X - e_2)(X - e_3).$$

Show that the cocycle

$$(\sigma\phi)\phi^{-1} = \theta_\sigma: \quad \mathcal{C} \to \mathcal{C}$$

for $\sigma \in \mathrm{Gal}(\kappa/\mathbf{Q})$ is of the type

$$\mathbf{x} \to \mathbf{x} + \mathbf{a}_\sigma$$

where $2\mathbf{a}_\sigma = \mathbf{o}$. Deduce that $\mathcal{C}$ is the jacobian of $\mathcal{D}$.

3. In this exercise the ground field is $\mathbf{Q}(\rho)$, where $\rho^3 = 1$, $\rho \neq 1$. Let a, b, $c \in \mathbf{Q}(\rho)$, and let

$$\mathcal{D}: \quad aU^3 + bV^3 + cW^3$$

$$\mathcal{C}: \quad X^3 + Y^3 + abcZ^3 = 0.$$

Put $\kappa = \mathbf{Q}(\rho, a^{1/3}, b^{1/3})$ and let $\phi: \mathcal{D} \to \mathcal{C}$ be given by

$$X = a^{1/3}U, \qquad Y = b^{1/3}V, \qquad Z = a^{-1/3}b^{-1/3}W.$$

Show that the corresponding θ_σ is

$$\theta_\sigma: \quad \mathbf{x} \to \mathbf{x}, \quad \text{or} \quad \mathbf{x} + (\rho, -\rho^2, 0) \quad \text{or} \quad \mathbf{x} + (\rho^2, -\rho, 0).$$

Deduce that $\mathcal{C}$ is the jacobian of $\mathcal{D}$.

The remaining exercises fill in the proof that (in the notation of the text) one can arrange to have $\varepsilon_\sigma = 1$ when $AB = 0$.

4. Let κ/k be a finite normal (separable) extension of fields of degree n. Let $\alpha_1, \ldots, \alpha_n$ be a basis of κ/k and let $\sigma_1, \ldots, \sigma_n$ be the elements of the Galois group. Show that

$$\det[\sigma_i \alpha_j] \neq 0.$$

[*Hint.* $\kappa = k(\beta)$ for some β.

Note. In what is still the finest introduction to Galois theory, (*Galois Theory.* Notre Dame Mathematical Lectures 2, 1942. Second edn., 1948.) E. Artin proves this at the onset by an induction argument.]

5. Let κ/k be a finite normal (separable) extension. For $\sigma \in \mathrm{Gal}(\kappa/k)$ let $\theta_\sigma \in k^*$ be given satisfying the cocycle identity

$$\theta_{\tau\sigma} = (\tau\theta_\sigma)\theta_\tau.$$

Show that $\{\theta_\sigma\}$ is a coboundary, i.e. that

$$\theta_\sigma = (\sigma\gamma)\gamma^{-1} \qquad (\text{all } \sigma),$$

for some $\gamma \in \kappa^*$.

[*Hint.* Let $\lambda \in \kappa$. Show that

$$\gamma = \sum_\sigma \theta_\sigma(\sigma\lambda)$$

does what is required provided that $\gamma \neq 0$. Use Lemma 2 to show that λ can be so chosen.

Note. This result is usually known as Hilbert 90 because it is Satz 90 in Hilbert's *Zahlbericht* - his report on algebraic number theory to the German Mathematical Society at the end of the last century.]

6. Let $n > 1$ be an integer. For $\sigma \in \mathrm{Gal}(\overline{\mathbf{Q}}/\mathbf{Q})$ let θ_σ be an nth root of 1 and suppose that $\{\theta_\sigma\}$ is a cocycle. Show that there is a $\delta \in \overline{\mathbf{Q}}$ such that $\theta_\sigma = \sigma\delta/\delta$ and $\delta^n \in \mathbf{Q}$.

21

Some abstract nonsense[25]

Let Γ be a *finite* group which acts on an *abelian* group A (written additively). The action is written σA ($\sigma \in \Gamma$, $a \in A$).

A *cocycle* is a map $\Gamma \to A$, say

$$\sigma \to a_\sigma$$

which satisfies the *cocycle identity*

$$\tau a_\sigma = a_{\tau\sigma} - a_\tau \qquad (\sigma, \tau \in \Gamma).$$

Note that for $\tau = 1$ (the identity of Γ) this implies

$$a_1 = 0.$$

If $b \in A$, then it is easy to see that

$$a_\sigma = \sigma b - b$$

is a cocycle. Cocycles of this type are called *coboundaries*.

Cocycles form a group under elementwise addition

$$\{a_\sigma\} + \{b_\sigma\} = \{a_\sigma + b_\sigma\}.$$

The coboundaries are a subgroup. The quotient group is

$$H^1(\Gamma, A),$$

---

[25] This is a self-contained account of what is needed from the cohomology of groups and commutative Galois cohomology. For how it fits into a wider picture, see, for example, Chapters IV and V of J.W.S. Cassels and A. Fröhlich (Editors) *Algebraic number theory*, Academic Press (1967). The treatment here is suggested by that in C. Chevalley *Class field theory*, Nagoya (1954).

the *first cohomology group.*

Now Γ acts on the whole situation ("transfer of structure"). Γ acts on itself by inner automorphisms. So τ acts on the map (cocycle)

$$\{a_\sigma\}: \quad \sigma \to a_\sigma$$

to give

$$\tau\{a_\sigma\}: \quad \tau\sigma\tau^{-1} \to \tau a_\sigma$$
$$= a_{\tau\sigma} - a_\tau.$$

Or, writing σ for $\tau\sigma\tau^{-1}$,

$$\tau(a_\sigma): \quad \sigma \to a_{\sigma\tau} - a_\tau.$$

This *is* a cocycle, as it has to be; and indeed

$$\tau\{a_\sigma\} - \{a_\sigma\}: \quad \sigma \to a_{\sigma\tau} - a_\tau - a_\sigma$$
$$= \sigma a_\tau - a_\tau$$

is a coboundary. Hence

Lemma 0. Γ *acts trivially on* $H^1(\Gamma, A)$.

Lemma 1. *Every element of* $H^1(\Gamma, A)$ *is of finite order dividing*[26] $\sharp\Gamma$.

Proof. Let the element be represented by the cocycle $\{a_\sigma\}$. Then, from what we have seen, it is also represented by the cocycle

$$\tau\{a_\sigma\} = \{a_{\sigma\tau} - a_\tau\}.$$

But now

$$\sum_\tau \tau\{a_\sigma\} = \{0\}$$

[Recall that $a_1 = 0$.]

Lemma 2. *Let* $m \in \mathbf{Z}$, $m > 1$. *Denote by* $\Delta_m \subset A$ *the set of elements of order dividing* m. *Suppose that every element of* A *is divisible by* m *in* A.

Then every element of $H^1(\Gamma, A)$ *of order* m *is representable by a cocycle* $\{d_\sigma\}$, $d_\sigma \in \Delta_m$.

Proof. Let the given element of $H^1(\Gamma, A)$ be represented by $\{a_\sigma\}$. By hypothesis, $m\{a_\sigma\}$ is a coboundary, say

$$ma_\sigma = \sigma b - b \quad (b \in A).$$

Under the hypotheses of the Lemma, $b = mc$, $c \in A$ so

$$ma_\sigma = m\sigma c - mc$$

---

[26] We use $\sharp$ for the cardinality of a set.

that is

$$m(a_\sigma - \sigma c + c) = 0.$$

Hence the element of H^1 is represented by

$$\sigma \to a_\sigma - \sigma c + c \in \Delta_m,$$

as required.

Denote by A^Γ the set of elements of A fixed by Γ:

$$a \in A^\Gamma \Leftrightarrow \sigma a = a \qquad \text{(all } \sigma \in \Gamma).$$

Lemma 3. *Notation and hypotheses as in previous Lemma. Then*

$$A^\Gamma / m A^\Gamma$$

is canonically isomorphic to a subgroup of $H^1(\Gamma, \Delta_m)$.

Proof. Let $a \in A^\Gamma$. By hypothesis

$$a = mb \qquad b \in A.$$

On applying $\sigma \in \Gamma$, we have

$$a = \sigma a = m \sigma b,$$

and so

$$m d_\sigma = 0, \qquad d_\sigma = \sigma b - b.$$

Hence $\{d_\sigma\}$ is a cocycle with values in Δ_m (indeed it becomes a coboundary in A).

For given a, any other choice of b is of the type $b + c$, $c \in \Delta_m$. Hence the element of $H^1(\Gamma, \Delta_m)$ given by $\{d_\sigma\}$ is uniquely determined by a.

If $a \in m A^\Gamma$, we may take $b \in A^\Gamma$, so $d_\sigma = 0$ for all σ, and the image in $H^1(\Gamma, \Delta_m)$ is 0.

Conversely suppose that the cocycle constructed above is a coboundary, so

$$d_\sigma = \sigma e - e \qquad \forall \sigma \in \Gamma, \text{ some } e \in \Delta_m$$

Then

$$\sigma(b - e) = b - e \qquad \text{(all } \sigma \in \Gamma):$$

and so

$$b - e \in A^\Gamma, \qquad m(b - e) = a.$$

We can put the last two lemmas together. We repeat the hypothesis.

Theorem. *Suppose that $m > 1$ is an integer and that every element of A is divisible by m. Then the sequence*

$$0 \to A^\Gamma / m A^\Gamma \to H^1(\Gamma, \Delta_m) \to [H^1(\Gamma, A)]_m \to 0$$

is exact, where [. . .]$_m$ *denotes the group of elements of order dividing* m, *and the third map is induced by* $\Delta_m \hookrightarrow A$.

Proof. After Lemmas 2, 3 we need only prove exactness at $H^1(\Gamma, \Delta_m)$, i.e. that the image of

$$A^\Gamma / mA^\Gamma \to H^1(\Gamma, \Delta_m)$$

is exactly the kernel of

$$H^1(\Gamma, \Delta_m) \to [H^1(\Gamma, A)]_m.$$

Consider first an element of the image, given (say) by the cocycle $\{d_\sigma\}$ By hypothesis, $d_\sigma = \sigma b - b$, $b \in A$ and so $\{d_\sigma\}$ considered as taking values in A, is a coboundary. Thus Image $\subset$ Kernel.

Now let the cocycle represented by $\{d_\sigma\}$ be in the kernel, i.e. $\{d_\sigma\}$ is a coboundary for A: $d_\sigma = \sigma b - b$ some $b \in A$. Then

$$\sigma(mb) - mb = md_\sigma = 0 \qquad \text{(all } \sigma\text{)}$$

and so $mb \in A^\Gamma$. Hence Kernel $\subset$ Image.

Galois cohomology. Let k be a field and $\overline{k}$ its separable closure (= algebraic closure in characteristic 0, the case of interest). Put

$$\Gamma = \mathrm{Gal}(\overline{k}/k).$$

We say that the action $a \to \sigma a$ ($\sigma \in \Gamma$, $a \in A$) of Γ on the abelian group A is *continuous* if:

For every $a \in A$ there is an extension κ of k of finite degree $[\kappa : k] < \infty$ (depending on a) such that

$$\sigma a = a \qquad \text{(all } \sigma \in \mathrm{Gal}(\overline{k}/\kappa) \subset \mathrm{Gal}(\overline{k}/k)\text{)}.$$

Note 1. An example is: $k = \mathbf{Q}$, C a curve $Y^2 = X^3 + AX + B$ defined over $\mathbf{Q}$, $A = \overline{\mho}$.

Note 2. If A has any natural topology, this is disregarded. For us the word "continuous" is just a term of art. The action is continuous in the usual sense if Γ is given an appropriate topology and A the *discrete* topology.

A *continuous cocycle* is a map

$$\sigma \to a_\sigma \qquad (\sigma \in \Gamma, a_\sigma \in A)$$

which

(i) satisfies the cocycle identity

$$\tau a_\sigma = a_{\tau\sigma} - a_\tau \qquad (\sigma, \tau \in \Gamma)$$

(ii) is continuous in the sense that there is a normal extension κ/k of finite degree $[\kappa : k] < \infty$ such that a_σ depends only on the action of σ on κ [Of course κ may depend on $\{a_\sigma\}$].

In particular,

$$a_\tau = 0 \qquad (\text{all } \tau \in \text{Gal}(\overline{k}/\kappa)),$$

so

$$\tau a_\sigma = a_{\tau\sigma} - a_\tau = a_\sigma - 0 \qquad (\text{all } \tau \in \text{Gal}(\overline{k}/\kappa)$$

and hence

$$a_\sigma \in \kappa \qquad (\text{all } \sigma \text{ in } \text{Gal}\,\overline{k}/k).$$

If $\{a_\sigma\}$, $\{b_\sigma\}$ are continuous cocycles, then clearly $\{a_\sigma + b_\sigma\}$ is continuous.

A coboundary $\{\sigma c - c\}$ $c \in A$ is automatically continuous, by our hypothesis that Γ acts continuously on A.

Definition. $H^1(\Gamma, A)$ *is the group of continuous cocycles modulo coboundaries.*

By following the proofs of the Γ finite case it is straightforward to prove

Theorem 1. $H^1(\Gamma, A)$ *is torsion (i.e. every element has finite order).*

Theorem 2. *Let $m > 1$ be an integer and suppose that every element of A is divisible by m. Then the sequence*

$$0 \to A^\Gamma/mA^\Gamma \to H^1(\Gamma, \triangle_m) \to [H^1(\Gamma, A)]_m \to 0$$

is exact where (as in the previous section)

(i) A^Γ *is the set of $a \in A$ fixed by Γ.*
(ii) $\triangle_m$ *is the set of elements of A of order dividing m.*
(iii) $[H^1(\Gamma, A)]_m$ *is the set of elements of $H^1(\Gamma, A)$ of order dividing m.*

Appendix.[27] Localization

Let p be a fixed prime. Choose a fixed embedding

$$\lambda : \overline{\mathbb{Q}} \hookrightarrow \overline{\mathbb{Q}}_p.$$

Write

[27] May be omitted at first reading. As will be explained, the result obtained here is obvious from another point of view in the context of the course.

$$\Gamma = \mathrm{Gal}(\overline{\mathbf{Q}}/\mathbf{Q})$$
$$\Gamma_p = \mathrm{Gal}(\overline{\mathbf{Q}}_p/\mathbf{Q}_p);$$

so λ induces an embedding

$$\lambda^* : \ \Gamma_p \hookrightarrow \Gamma.$$

Let A be a continuous Γ-module. Then it is via λ^* a continuous Γ_p-module.

Let $\{a_\sigma\}$, $\sigma \in \Gamma$ be a continuous cocycle. By restricting σ to Γ_p, we have a continuous Γ_p cocycle. Hence we have a group homomorphism

$$\lambda I : \ H^1(\Gamma, A) \to H^1(\Gamma_p, A)$$

[*localization*: A special case of the "restriction map"].

Ostensibly λI depends on the embedding λ, but we show that it does not.

Any embedding Λ of $\overline{\mathbf{Q}} \hookrightarrow \overline{\mathbf{Q}}_p$ is of the shape

$$\Lambda = \lambda\mu,$$

where μ is an automorphism of $\overline{Q}/Q$. By the analogue of Lemma 0 of the "Finite Γ" section, μ acts trivial on $H^1(\Gamma, A)$, and so $\Lambda I = \lambda I$. Thus the map

$$H^1(\Gamma, A) \to H^1(\Gamma_p, A)$$

is canonical.

In the context of the course, we have an elliptic curve

$$Y^2 = X^3 + AX + B$$

defined over $\mathbf{Q}$. Let $\overline{\mathcal{O}}$, $\overline{\mathcal{O}}_p$ be the points defined over $\overline{\mathbf{Q}}$, $\overline{\mathbf{Q}}_p$ respectively. We are concerned with the map

$$H^1(\Gamma, \overline{\mathcal{O}}) \to H^1(\Gamma_p, \overline{\mathcal{O}}_p),$$

which may be regarded as

$$H^1(\Gamma, \overline{\mathcal{O}}) \to H^1(\Gamma_p, \overline{\mathcal{O}}) \to H^1(\Gamma_p, \overline{\mathcal{O}}_p),$$

the second induced by the embedding

$$\overline{\mathcal{O}} \hookrightarrow \overline{\mathcal{O}}_p.$$

Later we interpret an element of $H^1(\Gamma, \overline{\mathcal{O}})$ as a curve $\mathcal{D}$ defined over $\mathbf{Q}$ together with a choice of structure as a principal homogeneous space.

A curve $\mathcal{D}$ defined over $\mathbf{Q}$ is certainly defined over $\mathbf{Q}_p$. with its structure of principle homogeneous space it thus corresponds to an element of $H^1(\Gamma_p, \overline{\mathcal{O}}_p)$. The resulting map $H^1(\Gamma, \overline{\mathcal{O}}) \to H^1(\Gamma_p, \overline{\mathcal{O}}_p)$ is precisely the one constructed above.

22

Principal homogeneous spaces and Galois cohomology

Let $\mathcal{D}$ be a curve of genus 1 defined over $\mathbf{Q}$. We have seen (§20, Theorem 1) that there is an elliptic curve

$$\mathcal{C}: \quad Y^2 = X^3 + AX + B$$

defined over $\mathbf{Q}$ and a birational equivalence

$$\phi: \quad \mathcal{D} \to \mathcal{C}$$

defined over $\overline{\mathbf{Q}}$. Further, for any $\sigma \in \mathrm{Gal}(\overline{\mathbf{Q}}/\mathbf{Q})$ the map

$$(\sigma\phi)\phi^{-1}: \quad \mathcal{C} \to \mathcal{C}$$

is of the type

$$\mathbf{x} \to \mathbf{x} + \mathbf{a}_\sigma,$$

where $\mathbf{a}_\sigma \in \overline{\mathfrak{G}}$.

The elliptic curve $\mathcal{C}$ is unique up to a transformation

$$X \to s^2 X, \qquad Y \to s^3 Y, \qquad s \in \mathbf{Q}^*.$$

Of course ϕ and the $\mathbf{a}_\sigma$ are far from being unique. $\mathcal{C}$ is the *jacobian* of $\mathcal{D}$.

We have to discuss how far the elements of the above situation are arbitrary. We note first that (by the previous discussion) the $\mathbf{a}_\sigma$ satisfy the cocycle identity

$$\tau \mathbf{a}_\sigma = \mathbf{a}_{\tau\sigma} - \mathbf{a}_\tau.$$

Now the $\mathbf{a}_\sigma$ are in the *commutative* group $\overline{\mathfrak{G}}$, and we may invoke the machinery of §21.

On replacing the map ϕ by $\phi\psi$. where

$$\psi: \ C \to C, \ \mathbf{x} \to \mathbf{x} + \mathbf{b} \ (\mathbf{b} \in \overline{\mathfrak{G}}),$$

we replace $\{\mathbf{a}_\sigma\}$ by

$$\mathbf{a}_\sigma + (\sigma\mathbf{b} - \mathbf{b})$$

where $\sigma\mathbf{b} - \mathbf{b}$ is a coboundary. In the commutative case, the coboundaries are a subgroup of the cocycles and so $\{a_r\}$ determines an element of the quotient group

$$\text{cocycles/coboundaries} \ = \ H^1(\Gamma, \overline{\mathfrak{G}})$$

- the first cohomology group, where $\Gamma = \text{Gal}(\overline{\mathbf{Q}}/\mathbf{Q})$.

We now look at the information which an element of $H^1(\Gamma, \overline{\mathfrak{G}})$ gives us about $\mathcal{D}$.

In the first place, we certainly can construct a curve $\mathcal{D}$ and a birational equivalence ϕ by our general machinery. To remind: let $\mathbf{x}$ be a generic point of C. There is an action $\tilde{\sigma}$ of $\text{Gal}(\overline{\mathbf{Q}}/\mathbf{Q})$ on $\overline{\mathbf{Q}}(\mathbf{x})$ given by

(i) $\tilde{\sigma}$ acts like σ on $\overline{\mathbf{Q}}$

(ii) $\tilde{\sigma}\mathbf{x} = \mathbf{x} + \mathbf{a}_\sigma$.

Then the fixed field is the function field of a curve $\mathcal{D}$ defined over $\mathbf{Q}$ and ϕ, defined over $\overline{\mathbf{Q}}$, is given by the identification of the two function fields over $\overline{\mathbf{Q}}$.

The map ϕ gives $\mathcal{D}$ a structure of *principal homogeneous space* over C in the following sense.

Let ξ_1, ξ_2 be independent generic points on $\mathcal{D}$, which we treat as fixed under $\text{Gal}(\overline{\mathbf{Q}}/\mathbf{Q})$. Put

$$\Delta(\xi_1, \xi_2) = \phi(\xi_1) - \phi(\xi_2).$$

Then

$$\sigma\Delta(\xi_1, \xi_2) = (\phi(\xi_1) + \mathbf{a}_\sigma - (\phi(\xi_2) + \mathbf{a}_\sigma)$$
$$= \Delta(\xi_1, \xi_2).$$

That is, the algebraic map from two copies of $\mathcal{D}$ to C given by Δ is defined over $\mathbf{Q}$. Clearly

$$\Delta(\xi_1, \xi_2) + \Delta(\xi_2, \xi_3) = \Delta(\xi_1, \xi_3).$$

Hence the cocycle $\{\mathbf{a}_\sigma\}$, or the corresponding elements of $H^1(\Gamma, \overline{\mathfrak{G}})$, determines the pair $(\mathcal{D}, \Delta)$. The cocycle $\{-\mathbf{a}_\alpha\}$ determines the pair $(\mathcal{D}, -\Delta)$. Thus to get a group structure we must consider not just the curves $\mathcal{D}$ with given jacobian, but the pairs $(\mathcal{D}, \Delta)$ where Δ is a structure of principal homogeneous space.

The above account overlooks one tricky point. An element of $H^1(\Gamma, \overline{\mathfrak{G}})$ determines the function field of $\mathcal{D}$, and so determines $\mathcal{D}$ only up to birational equivalence defined over $\mathbf{Q}$. Now it can happen that there is a birational automorphism of $\mathcal{D}$ defined over $\mathbf{Q}$ which interchanges Δ and $-\Delta$ (!). A trivial example is when $\mathcal{C}$ is regarded as its own jacobian. Consider two maps

$$\phi_j : \quad (\mathcal{D} =)\, \mathcal{C} \to \mathcal{C}, \qquad (j = 1, 2)$$

where ϕ_1 is $\mathbf{x} \to \mathbf{x}$ and ϕ_2 is $\mathbf{x} \to -\mathbf{x}$. In both cases the cocycle $\mathbf{a}_\sigma$ is identically 0. In the first case, $\Delta_1(\mathbf{x}_1, \mathbf{x}_2) = \mathbf{x}_1 - \mathbf{x}_2$; and in the second $\Delta_2(\mathbf{x}_1, \mathbf{x}_2) = \mathbf{x}_2 - \mathbf{x}_1$. The two are taken into one another by the automorphism $\mathbf{x} \to -\mathbf{x}$ of $\mathcal{C} = \mathcal{D}$.

In the example just above, we have the trivial element of $H^1(\Gamma, \overline{\mathfrak{G}})$. There is the same phenomenon for elements of order 2 (and only for them) [Exercise for reader!].

To deal with this difficulty, we shall identify two structures of principal homogeneous space which are birationally equivalent. With this convention each element of $H^1(\Gamma, \overline{\mathfrak{G}})$ defines a unique principal homogeneous space.

Conversely, a structure of principal homogeneous space determines the element of $H^1(\Gamma, \overline{\mathfrak{G}})$. Consider the map

$$\phi : \qquad \mathcal{D} \to \mathcal{C}.$$

By our initial construction, the corresponding cocycle is

$$\mathbf{a}_\sigma = (\sigma\phi)(\xi) = \phi(\xi),$$

where ξ is a generic point of $\mathcal{D}$ fixed under Galois.

Now let α be any algebraic point on $\mathcal{D}$ (i.e. defined over $\overline{\mathbf{Q}}$). Then

$$\sigma(\phi(\alpha)) = (\sigma\phi)(\sigma\alpha) = \phi(\sigma\alpha) + \mathbf{a}_\sigma,$$

since σ acts both α and on the coefficients of the map ϕ. Hence

$$\Delta(\alpha, \sigma\alpha) = \phi(\alpha) - \phi(\sigma\alpha)$$
$$= \phi(\alpha) - \sigma(\phi(\alpha)) + \mathbf{a}_\sigma$$

Thus $\{\Delta(\alpha, \sigma\alpha)\}_\sigma$ is a cocycle, and differs from $\{\mathbf{a}_\sigma\}_\sigma$ by a coboundary.

To sum up:

Theorem *There is a canonical isomorphism between principal homogeneous spaces* $(\mathcal{D}, \Delta)$ *(up to birational equivalence over $\mathbf{Q}$) and elements of $H^1(\Gamma, \overline{\mathfrak{G}})$. The element corresponding to $(\mathcal{D}, \Delta)$ is given by the cocycle $\{\Delta(\alpha, \sigma\alpha)\}_\sigma$, where α is any algebraic point on $\mathcal{D}$.*

Note 1. Principal homogeneous spaces were introduced by Weil. He defined their group structure directly, not by reference to $H^1(\Gamma, \overline{\mathfrak{G}})$.

Note 2. For the cognoscenti. The "jacobian" defined here is a refinement of the classical notion defined over the complex numbers.

Recall that a divisor $\mathfrak{a}$ on $\mathcal{D}$ is a map from the algebraic points α on $\mathcal{D}$ to $\mathbf{Z}$ which is 0 for all except at most finitely many α. It is defined over $\mathbf{Q}$ if it is **invariant** (in an obvious sense) under $\mathrm{Gal}(\overline{\mathbf{Q}}/\mathbf{Q})$. The degree is $\sum n_\alpha$, where $\mathfrak{a}$ is $\alpha \to n_\alpha$.

Suppose that $\mathfrak{a}$ is of degree 0. The jacobian map is the map from $\mathfrak{a}$ to

$$\mathrm{Jac}(\mathfrak{a}) = \sum_\alpha n_\alpha \phi(\alpha) \in \overline{\mathfrak{G}},$$

the summation being that on $\mathcal{C}$.

The divisor $\mathfrak{a}$ is in the kernel of the map precisely when the $\phi(\alpha)$ with their multiplicities are the poles and zeros of a function on $\mathcal{C}$. Identifying $\mathcal{D}$ and $\mathcal{C}$ via ϕ, this is the same as saying that $\mathfrak{a}$ is the divisor of a function on $\mathcal{D}$ [a *principal* divisor].

If $\mathfrak{a}$ is defined over $\mathbf{Q}$, then $\mathrm{Jac}(\mathfrak{a})$ is defined over $\mathbf{Q}$, as follows easily from the formula for $\sigma\phi(\alpha)$. Hence we have group monomorphism.

$$\frac{\text{Divisors of degree 0 on } \mathcal{D} \text{ defined over } \mathbf{Q}}{\text{Principal such divisors}} \to \mathfrak{G}.$$

A final point. If the divisor $\mathfrak{a}$ of degree 0 is defined over $\mathbf{Q}$ and is principal, then it is the divisor of a function on $\mathcal{D}$ defined over $\mathbf{Q}$. For suppose that f is a function with divisor $\mathcal{D}$ defined over $\overline{\mathbf{Q}}$. Let $\sigma \in \mathrm{Gal}(\overline{\mathbf{Q}}/\mathbf{Q})$. Then $\mathfrak{a}$ is also the divisor of σf and so

$$\frac{\sigma f}{f} \in \overline{\mathbf{Q}}^*.$$

It is readily checked that $\sigma \to \sigma f/f$ is a cocycle with values in $\overline{\mathbf{Q}}^*$; and so is a coboundary by Hilbert 90 [§20, Exercise 5]. Hence $\sigma f/f = \sigma\lambda/\lambda$ for some $\lambda \in \overline{\mathbf{Q}}^*$ and all σ. Then $\lambda^{-1} f$ is fixed under Galois, i.e. defined over $\mathbf{Q}$, and has divisor $\mathfrak{a}$, as required. [Of course this remark is general, and applies to curves of any genus.]

§22. Exercises

1. If ξ, $\mathbf{x}$ are generic points of $\mathcal{D}$, $\mathcal{C}$ respectively, fixed under Galois, show that the function $A(\xi, \mathbf{x}) = \phi^{-1}(\phi(\xi) + \mathbf{x})$ is defined over $\mathbf{Q}$ and investigate its properties.

The Tate-Shafarevich group

We put together the results of the two previous sections.
As before, let

$$C: \quad Y^2 = X^3 + AX + B$$

be an elliptic curve defined over $\mathbf{Q}$. The groups of points defined over $\mathbf{Q}$, $\overline{\mathbf{Q}}$ respectively are $\mathfrak{G}$, $\overline{\mathfrak{G}}$; and Γ is $\mathrm{Gal}(\overline{\mathbf{Q}}/\mathbf{Q})$. We have seen that the first cohomology group $H^1(\Gamma, \overline{\mathfrak{G}})$ is canonically isomorphic to the group of equivalence classes of $\{\mathcal{D}, \Delta\}$ where $\mathcal{D}$ is a curve of genus 1 and Δ is a structure of principal homogeneous space on it. This group is often referred to as the Weil-Châtelet group and denoted by $WC = WC(C)$.

Let $m > 1$ be an integer. The group $\overline{\mathfrak{G}}$ is divisible by m since finding a $\mathbf{b}$ such that $m\mathbf{b} = \mathbf{a} \in \overline{\mathfrak{G}}$ is just a matter of solving some algebraic equations. The exact sequence of the previous section is now

$$0 \to \mathfrak{G}/m\mathfrak{G} \to H^1(\Gamma, \Delta_m) \to [H^1(\Gamma, \overline{\mathfrak{G}})]_m \to 0,$$

where $\Delta_m \subset \overline{\mathfrak{G}}$ is the group of elements of $\overline{\mathfrak{G}}$ of order m and the $[\ldots]_m$ denotes the subgroup of elements of order dividing m.

We now have an approach to the weak Mordell-Weil theorem. We would like to find the elements of $H^1(\Gamma, \Delta_m)$ which are the images of $\mathfrak{G}/m\mathfrak{G}$. By the exactness of the sequence these are precisely the kernel of the map

$$H^1(\Gamma, \Delta_m) \to H^1(\Gamma, \overline{\mathfrak{G}}) = WC(C).$$

Being in the kernel means that the image is a trivial principal homogeneous space $\{\mathcal{D}, \Delta\}$; i.e. that there is a point on $\mathcal{D}$ defined over $\mathbf{Q}$.

For $m = 2$ we are back in the situation discussed in the proof of the Weak Mordell-Weil Theorem. There we displayed the curve $\mathcal{D}$ in the image $\{\mathcal{D}, \Delta\}$ of an element of $H^1(\Gamma, \Delta_2)$ as the intersection of two quadratic surfaces[28].

As we have already emphasised, there is even now no algorithm for deciding whether or not there is a rational point on $\mathcal{D}$. There is, however, no difficulty in deciding whether or not there is a point on $\mathcal{D}$ everywhere locally. As we shall see in a moment, the elements of WC for which there is a point on $\mathcal{D}$ everywhere locally form a subgroup. It is known as the Tate-Shafarevich group and is usually denoted[29] by the Russian letter Ш ("sha").

To show that Ш is a subgroup we must discuss localization. For any prime p (including ∞) we use a suffix p to denote an object defined over $\mathbf{Q}_p$ instead of over $\mathbf{Q}$. There is an obvious map

$$j_p : \quad WC \to WC_p$$

which takes the equivalence class of a principal homogeneous space $(\mathcal{D}, \Delta)$ defined over $\mathbf{Q}$ into the class of the same $\{\mathcal{D}, \Delta\}$ considered over $\mathbf{Q}_p$. The non-cohomological description of the composition of principal homogeneous spaces works entirely over the ground field: thus it shows immediately that the localization j_p respects the group law; but we have not explained that description. From the cohomological point of view, we have a map

$$j_p : \quad H^1(\Gamma, \overline{\mathfrak{G}}) \to H^1(\Gamma_p, \overline{\mathfrak{G}}_p),$$

induced by the inclusion $\overline{\mathfrak{G}} \subset \overline{\mathfrak{G}}_p$. This situation was discussed at the end of §21, where it was shown that j_p is a group homomorphism and is independent of the choice of inclusion $\overline{\mathbf{Q}} \subset \overline{\mathbf{Q}}_p$.

Clearly Ш is the intersection of the kernels of all the localization maps j_p (including $p = \infty$). For given m, denote by S_m the group of elements of $H^1(\Gamma, \Delta_m)$ which map into Ш $\subset H^1(\Gamma, \overline{\mathfrak{G}})$. It is called the mth Selmer group. Now we have the exact sequence

$$0 \to \mathfrak{G}/m\mathfrak{G} \to S_m \to [\,Ш\,]_m \to 0.$$

For $m = 2$, which we encountered in the proof of the weak Mordell-Weil Theorem, we saw that S_2 is finite and effectively constructible. It

---

[28] The author apologizes for the clash between Δ denoting a structure of principal homogeneous space and Δ_2, the group of elements of order 2 in $\overline{\mathfrak{G}}$.

[29] This is the author's most lasting contribution to the subject. The original notation was TS, which, Tate tells me, was intended to continue the lavatorial allusion of WC. The Americanism "tough shit" indicates the part that is difficult to eliminate.

can be shown by a more sophisticated version of the same argument that
the same things hold for S_m and general m, though now the effective
constructibility tends to be not very practical.

To sum up. The Selmer group is knowable. It majorizes $\mho/m\mho$ and
the "error" is given by Ш, which can be called the obstruction to the
local-glocal principle for curves of genus 1 with the given jacobian $\mathcal{C}$.

This is as far as we shall go in this direction with the theory. We
conclude with background comments.

Before all this theory was invented, Selmer embarked on a massive
programme to find the Mordell-Weil groups of elliptic curves, especially
those of the type

$$\mathcal{C}: \quad X^3 + Y^3 + dZ^3 = 0,$$

where $d \in \mathbf{Z}$. He used descent arguments to bound the Mordell-Weil
rank. Also, by a direct search, he found rational points on $\mathcal{C}$ and so
bounded the Mordell-Weil rank from below. Most often the upper and
the lower estimates for the rank coincided, but when there was a dis-
crepancy the difference was always even. Moreover, estimates for the
rank derived from different types of descent (e.g. majorization of $\mho/2\mho$
and $\mho/3\mho$) always differed, if at all, by an even integer.

After the group Ш was discovered by Tate and Shafarevich, it was
natural to look for the explanation of this phenomenon in the structure
of Ш. It turns out that there is a skew-symmetric form on Ш whose
kernel is the group of infinitely-divisible elements of Ш. It always
looked improbable that there are infinitely-divisible elements and by now
there is much evidence (but no proof) that they do not exist. If there
are no infinitely-divisible elements, the existence of the skew-symmetric
form shows that the order of $[\,\text{Ш}\,]_m$ is a square. This explains Selmer's
observation.

There is not merely a local-global principle for curves of genus 0, but
it has a quantitative formulation (and also, more generally for linear algebraic groups. The modern formulation is in terms of the "Tamagawa
number"). On the basis of massive calculations (this time on a computer) Birch and Swinnerton-Dyer proposed what can be regarded as
a quantitative local-global theorem for elliptic curves. In their formula
there is a number, not otherwise accounted for. In all their calculations
the mysterious number turned out to be an integer and indeed a perfect
square.

It was natural to interpret this integer as the order of Ш (supposed finite), and, once made, this identification was supported on other grounds.

The Birch-Swinnerton-Dyer conjectures were widely generalized and further evidence for their plausibility were adduced. It is only in the last few years that progress has been made with their proof. Until the very recent work of Rubin and Kolyvagin there was not even a single elliptic curve for which Ⅲ had been proved to be finite.

§23. Exercises

1. Let m, n be integers, $m \mid n$. Show that there is a group homomorphism λ such that

$$
\begin{array}{ccc}
\mathfrak{G} & \longrightarrow & H^1(\Gamma, \Delta_m) \\
& \searrow & \downarrow \lambda \\
& & H^1(\Gamma, \Delta_n)
\end{array}
$$

commutes.

Hence show that there are μ, ν such that

$$
\begin{array}{ccccccccc}
0 & \longrightarrow & \mathfrak{G}/n\mathfrak{G} & \longrightarrow & H^1(\Gamma, \Delta_n) & \longrightarrow & [WC]_n & \longrightarrow & 0 \\
& & \downarrow \mu & & \downarrow \lambda & & \downarrow \nu & & \\
0 & \longrightarrow & \mathfrak{G}/m\mathfrak{G} & \longrightarrow & H^1(\Gamma, \Delta_m) & \longrightarrow & [WC]_m & \longrightarrow & 0
\end{array}
$$

is exact and commutative.

Describe μ, ν explicitly.

24

The endomorphism ring

In this section, the ground field k is any field, possibly of characteristic $p \neq 2, 3$. [This last restriction solely because of our choice of canonical form.] The main objective is the application to the estimation of the number of points over finite fields, but we do a little more, to set things in context.

Let

$$C: \quad Y^2 = X^3 + AX + B$$

be an elliptic curve defined over k. An endomorphism of C (over k) is a rational map

$$\phi: \quad C \to C$$

defined over k, for which

$$\phi(o) = o.$$

One endomorphism is the constant isomorphism which maps C entirely onto o. Otherwise, if $\mathbf{x}$ is a generic point of C, then so is

$$\xi = \phi(\mathbf{x})$$

and $k(\mathbf{x})/k(\xi)$ is an algebraic extension. We define the degree of ϕ to be

$$d(\phi) = [k(\mathbf{x}) : k(\xi)].$$

By convention, the degree of the constant endomorphism is 0.

The first lemma shows that ϕ respects the group structure of C. It is not really needed for what follows, but it helps to set ideas. In the application to finite fields, the conclusion will be obvious.

Lemma 1. *Let* **a**, **b** *be points of* C. *Then*

$$\phi(\mathbf{a} + \mathbf{b}) = \phi(\mathbf{a}) + \phi(\mathbf{b}).$$

Sketch proof. By extending the ground field if necessary, we may suppose that **a**, **b** are defined over k. If ϕ is the constant endomorphism, there is nothing to prove. Otherwise, let **x** be a generic point $\boldsymbol{\xi} = \phi(\mathbf{x})$. By the definition of the group law, there is a

$$\lambda = \lambda(\mathbf{x}) \in k(\mathbf{x})$$

whose only zeros are simple zeros at **a**, **b** and whose only poles are simple poles at **o**, **a** + **b**. Let

$$\Lambda = \Lambda(\boldsymbol{\xi}) = \mathrm{Norm}_{k(\boldsymbol{\xi})/k(\mathbf{x})}\, \lambda.$$

Then the zeros of Λ are just simple zeros at $\phi(\mathbf{a})$, $\phi(\mathbf{b})$ and the poles of Λ are just simple poles at $\phi(\mathbf{a} + \mathbf{b})$ and at $\mathbf{o} = \phi(\mathbf{o})$.

Note. cf. §14, Lemma 1. The proof above follows that in Silverman, Theorem 4.8 (p. 75), where it is proved for isogenies and the treatment is fuller. For the corresponding result for abelian varieties of any dimension, see D. Mumford, *Abelian Varieties* (Oxford, 1970), p.43, Corollary 3 or H.P.F. Swinnerton-Dyer, *Analytic theory of abelian varieties* (Cambridge, 1974), Theorem 32 or S. Lang, *Abelian varieties* (New York and London, 1959), Chapter II, Theorem 4.

All we shall need is the

Corollary. *Let* **x**, $\boldsymbol{\xi}$ *be as above and let*

$$\mathbf{x} = (x, y), \qquad \boldsymbol{\xi} = (\xi, \eta).$$

Then $\xi \in k(x)$, *and*

$$[k(x) : k(\xi)] = [k(\mathbf{x}) : k(\boldsymbol{\xi})] = d(\phi).$$

Proof. For $\phi(-\mathbf{x}) = -\phi(\mathbf{x}) = -\boldsymbol{\xi}$.

For any two endomorphisms ϕ, ψ, we defined the sum $\phi + \psi$ and the product $\phi\psi$ by

$$(\phi + \psi)(\mathbf{x}) = \phi(\mathbf{x}) + \psi(\mathbf{x}),$$
$$(\phi\psi)(\mathbf{x}) = \phi(\psi(\mathbf{x})),$$

where **x** is a generic point. It is readily verified that this gives the set of endomorphisms the structure of a (not necessarily commutative) ring.

Lemma 2.
$$d(\phi\psi) = d(\phi)d(\psi).$$

Proof. Clear.

Lemma 3.

$$d(\phi + \psi) + d(\phi - \psi) = 2d(\phi) + 2d(\psi).$$

Proof. Let $\mathbf{x} = (x, y)$ be a generic point, and put

$$\phi(\mathbf{x}) = \xi_1,$$
$$\psi(\mathbf{x}) = \xi_2,$$
$$(\phi + \psi)(\mathbf{x}) = \xi_3,$$
$$(\phi - \psi)(\mathbf{x}) = \xi_4,$$

so

$$\xi_3 = \xi_1 + \xi_2, \qquad \xi_4 = \xi_1 - \xi_2.$$

Then

$$\xi_j \in k(x), \qquad (j = 1, 2, 3, 4)$$

where

$$\xi_j = (\xi_j, \eta_j).$$

We argue as in the corresponding results for heights (§17, Lemma 4). The degree of an element of $k(x)$ corresponds to the height of an element of $\mathbf{Q}$. As $k(x)$ has no archimidean valuations trivial on k, the results are more precise.

By the formula for sum and difference, we have

$$1 : \xi_3 + \xi_4 : \xi_3\xi_4 =$$
$$(\xi_1 - \xi_2)^2 : 2(\xi_1\xi_2 + A)(\xi_1 + \xi_2) + 4B$$
$$: \xi_1^2\xi_2^2 - 2A\xi_1\xi_2 - 4B(\xi_1 + \xi_2) + A^2.$$

A similar argument to that for heights[30] gives

$$\deg(\xi_3) + \deg(\xi_4) = 2\deg(\xi_1) + 2\deg(\xi_2),$$

where "deg" is the degree as a rational function of x (= maximum of the degrees of numerator and denominator.)

This result now follows from Lemma 1, Corollary.

---

[30] cf. also (*) of §17

Corollary. *There are* r, s, $t \in \mathbf{Z}$, *depending on* ϕ, ψ, *such that*
$$d(m\phi + n\psi) = rm^2 + smn + tn^2$$
for all m, $n \in \mathbf{Z}$.
   Further,
$$r \geq 0, \qquad t \geq 0, \qquad s^2 - 4rt \geq 0.$$

Proof. The first part follows exactly as for heights.[31] For the second, $d(.) \geq 0$ by definition, so the quadratic form in m, n is positive semi-definitive or definite.

The rest of this section is not needed for the application to finite fields.

By abuse of notation we denote the constant endomorphism by 0 and the identity endomorphism $\phi(\mathbf{x}) = \mathbf{x}$ by 1.

Lemma 4. *Every endomorphism* ϕ *satisfies a quadratic equation*
$$\phi^2 - s\phi + t = 0,$$
where s, $t \in \mathbf{Z}$.

Proof. By the preceding Lemma,
$$d(m + n\phi) = m^2 + smn + tn^2$$
for some s, $t \in \mathbf{Z}$ and for all m, $n \in \mathbf{Z}$.
   Let $l \in \mathbf{Z}$. Then
$$d(\phi + l) = d(\phi - s - l) = l^2 + sl + t.$$
Hence by Lemma 2
$$d((\phi + l)(\phi - s - l)) = (l^2 + sl + t)^2. \tag{$*$}$$
But
$$(\phi + l)(\phi - s - l) = \phi^2 - s\phi - l(s + l). \tag{$**$}$$
Hence and by Lemma 3, Corollary, with $\phi^2 - s\phi$, 1 for ϕ, ψ, we have
$$d(\phi^2 - s\phi + n) = (-n + t)^2$$
for all $n \in \mathbf{Z}$. In particular,
$$d(\phi^2 - s\phi + t) = 0.$$
But the only endomorphism of degree 0 is the constant endomorphism 0.

---

[31] cf. §17, Exercise 2.

Note. As was shown by Deuring, the endomorphism ring is isomorphic to one of:

(i) $\mathbf{Z}$.

(ii)   a ring of integers in an imaginary quadratic field

(iii) a ring of integers in a generalized quaternion skew field.

   The last case can occur only in characteristic $p \neq 0$; and the skew field is very special.

§24 Exercises

1. Suppose that the ground field contains an element i with $i^2 = -1$ and that its characteristic is not 2. Let C be $Y^2 = X^3 + AX$ for some $A \neq 0$. Show that

$$\varepsilon : \quad Y \to iY, \qquad X \to X$$

is an endomorphism.

   Construct the endomorphism $1 + \varepsilon$ and check that $(1 + \varepsilon)^2 = 2\varepsilon$ as endomorphisms.

2. Suppose that the characteristic of the ground field is not 2 or 3 and that it contains ρ with $\rho^3 = 1$, $\rho \neq 1$. Let C be $Y^2 = X^3 + B$ for some $B \neq 0$. Show that

$$\lambda : \quad X \to \rho X, \qquad Y \to Y$$

is an endomorphism. Construct the endomorphism $\lambda - \lambda^2$ and show that $(\lambda - \lambda^2)^2 = -3$ as endomorphisms.

3. Suppose that the characteristic of the ground field is not 2. For $a \neq 0$ determine the b such that the isogenous curves

$$C : \quad Y^2 = X(X^2 + aX + b)$$
$$C_1 : \quad Y^2 = X(X^2 - 2aX + a^2 - 4b)$$

are birationally equivalent over the algebraic closure. Show that they are equivalent over the ground field provided that -2 is a square in it.

   Denote the isogeny, considered as an endomorphism of C, by μ. Show that $\mu^2 = -2$ as endomorphisms.

4. Let

$$\phi : C \to C$$

be an endomorphism and suppose that

$$\phi^2 - s\phi + t = 0 \qquad s, t \in \mathbf{Z}.$$

For positive integer m show that $\psi = \phi^m$ satisfies

$$\psi^2 - s_m\psi + t_m = 0,$$

where $s_m, t_m \in \mathbf{Z}$ are defined as follows. Let $\alpha, \beta \in \overline{\mathbf{Q}}$ be the roots of $T^2 - sT + t = 0$. Then

$$s_m = \alpha^m + \beta^m, \qquad t_m = \alpha^m\beta^m.$$

5. (i) Let ϕ be an endomorphism and define ϕ' by

$$\phi' = \phi \quad \text{if} \quad \phi \in \mathbf{Z};$$

otherwise $\phi' = s - \phi$, where $\phi^2 - s\phi + t = 0$. Show that

$$\phi\phi' = \phi'\phi = d(\phi).$$

(ii) Let $\mathbf{x}$ be a generic point and let $\xi_1, \ldots, \xi_t$ be the points of C defined over $\overline{k(\mathbf{x})}$ (k = ground field) such that $\phi(\xi_j) = \mathbf{x}$ (with appropriate multiplicities if ϕ is inseparable). Show that

$$\phi'(\mathbf{x}) = \sum \mathbf{x}_j$$

(addition on C).

(iii) If ψ is another endomorphism, show that

$$(\phi\psi)' = \psi'\phi'$$

and

$$(\phi + \psi)' = \phi + \psi'.$$

6. Let $K = k(t)$, where t is transcendental over k. Suppose that the elliptic curve C is defined over k. If $\xi = (\xi, \eta)$ on C is defined over K, show that it is already defined over k.

[*Hint.* Define s by $s^2 = F(t)$, where C is $Y^2 = F(X)$. Apply Lemma 1 to the map $(t, s) \to (\xi, \eta)$ of $C \to C$.

Note. cf. Miles Reid *Undergraduate Algebraic Geometry* (LMS Student Texts 12), p. 28.

Exercise. Motivate Reid's proof.]

25

Points over finite fields

We denote by $\mathbf{F}_q$ the field of q elements and denote its characteristic by p, so q is a power of p. Our objective is the

Theorem 1. *Let*

$$C: \quad Y^2 = X^3 + AX + B$$

be an elliptic curve over a finite field $\mathbf{F}_q$. The number N of points of C defined over $\mathbf{F}_q$ satisfies

$$|N - (q+1)| \le 2q^{1/2}.$$

We shall give the main idea of a proof but will have to be impressionist on one of the ingredients. Because of our canonical form, we shall assume that $p \ne 2, 3$. Note that N includes the point $\mathbf{o}$ "at infinity". At the end of the section we shall indicate the proof of a couple of other results.

Let $\mathbf{x} = (x, y)$ be a generic point. We show that $\phi(\mathbf{x}) = (x^q, y^q)$ is also on the curve. Indeed, since we are in characteristic $p \mid q$,

$$(y^q)^2 = (x^3 + Ax + B)^q$$
$$= (x^q)^3 + A^q x^q + B^q$$
$$= (x^q)^3 + Ax^q + B,$$

as $A^q = A$, $B^q = B$. This is the *Frobenius endomorphism*.

Now let $\mathbf{u} = (u, v)$ be a point defined over the algebraic closure $\overline{\mathbf{F}}_p$. Then

$$\phi(\mathbf{u}) = (u^q, v^q),$$

so $\mathbf{u}$ is defined over $\mathbf{F}_q$ precisely when it is a fixed point of ϕ or, what is the same thing, when

$$(\phi - 1)\mathbf{u} = \mathbf{o},$$

where 1 is the identity endomorphism and $\phi - 1$ is defined in terms of the endomorphism ring.

In the notation of the previous section, clearly

$$d(\phi) = q$$

and so by §24, Lemma 3, Corollary

$$d(\phi - 1) = q - s + 1$$

where

$$s^2 \leq 4q, \qquad |s| \leq 2q^{1/2}.$$

We have seen that a point defined over $\overline{\mathbf{F}}_q$ is actually defined over $\mathbf{F}_q$ precisely when it is the kernel of $\phi - 1$. But the degree of an endomorphism is equal to the number of algebraic points in the kernel, each counted with its multiplicity. If therefore we can show that the points of the kernel of $\phi - 1$ have multiplicity 1, we are done.

It is here that we have to leave a lacuna. One argument, which can be made precise, is to observe that $dx^q/dx = qx^{q-1} = 0$ in characteristic p, and so the differential of the map $\phi - 1$ is the same as that of the map -1, and hence never 0.

Note. The result is due to Hasse by essentially the same proof. It is often referred to as the "Riemann hypothesis for function fields" (of genus 1) because of an analogy with Riemann's notorious unproved conjecture about the zeros of the usual ("Riemann") zeta function. It was generalized to curves of any genus by Weil and to algebraic varieties by Deligne. The analysis of the action of the Frobenius map ϕ is still a central theme of modern arithmetic geometry.

Theorem 2. *Let $\mathcal{D}$ be a curve of genus 1 defined over $\mathbf{F}_q$. Then it has a point defined over $\mathbf{F}_q$.*

Proof. We developed the theory of the jacobian in characteristic 0, but it holds for general characteristics. Let $\mathcal{C}$ be the jacobian of $\mathcal{D}$ and let $\overline{\mathfrak{G}}$ be the group of points on $\mathcal{C}$ defined over $\overline{\mathbf{F}}_q$. It is enough to show that

$$H^1(\Gamma, \overline{\mathfrak{G}})$$

is trivial, where

$$\Gamma = \mathrm{Gal}(\overline{\mathbf{F}}_q/\mathbf{F}_q).$$

The group Γ is generated[32] by the Frobenius automorphism γ (say): $a \to a^q$. We have to show that any cocycle $\{a_o\}$ is trivial. It is enough to show that

$$\mathbf{a}_\gamma = \gamma\mathbf{b} - \mathbf{b}$$

for some $\mathbf{b} \in \overline{\mathfrak{G}}$. Now

$$\gamma\mathbf{b} - \mathbf{b} = (\phi - 1)\mathbf{b}$$

where ϕ is the "geometrical" Frobenius, so $\phi - 1$ is not the constant endomorphism. For any $\mathbf{c} \in \overline{\mathfrak{G}}$ we can thus solve $(\phi - 1)\mathbf{b} = \mathbf{c}$ for $\mathbf{b}$, since we are working in the algebraic closed field. In particular, this holds for $\mathbf{c} = \mathbf{a}_\gamma$. The cocycle identity gives inductively that

$$\mathbf{a}_\sigma = \sigma\mathbf{b} - \mathbf{b} \qquad \sigma = \gamma, \gamma^2, \gamma^3, \ldots$$

and we are done.

Note. For a broad generalization, see S. Lang, Algebraic groups over finite fields. *Amer. J. Math.* **78** (1956), 535-563.

The Theorem is due to F.K. Schmidt and the idea behind his proof is amusing. He used analytic means to estimate the number of points defined over the extension fields $\mathbf{F}_{q^n}$. In particular, he showed that the number is > 0 for all large enough n.

Let $\mathbf{b}_1, \ldots, \mathbf{b}_n$ be n conjugate points defined over $\mathbf{F}_{q^n}$ and $\mathbf{c}_1, \ldots, \mathbf{c}_{n+1}$ be similar conjugates defined over $\mathbf{F}_{q^{n+1}}$. Then by Riemann-Roch there is a function whose poles are simple poles at the $\mathbf{c}_i$ and which has simple zeros at the $\mathbf{b}_j$. It has one further zero; which must be defined over $\mathbf{F}_q$.

Theorem 3. *Let*

$$\lambda: \quad \mathcal{C}_1 \to \mathcal{C}_2$$

be an isogeny of elliptic curves, everything defined over $\mathbf{F}_q$. *Then* $N_1 = N_2$, *where* N_j *is the number of points on* $\mathcal{C}_j$ *defined over* $\mathbf{F}_q$.

Note. An isogeny is defined to be a rational map onto such that $\lambda(\mathbf{o}_1) = \mathbf{o}_2$. Lemma 1 of the preceding section extends to isogenies, which gives compatibility with the usage earlier in the course.

Proof. Let ϕ_j be the Frobenius on $\mathcal{C}_j$. Clearly the diagram

---

[32] "topologically", that is, γ generates the Galois group of every finite normal extension.

$$C_1 \xrightarrow{\phi_1} C_1$$
$$\downarrow \lambda \qquad \downarrow \lambda$$
$$C_2 \xrightarrow{\phi_2} C_2$$

is commutative, and hence so is

$$C_1 \xrightarrow{\phi_1 - 1} C_1$$
$$\downarrow \lambda \qquad \downarrow \lambda$$
$$C_2 \xrightarrow{\phi_2 - 2} C_2$$

It follows that the degrees

$$d(\phi_1 - 1) = d(\phi_2 - 1)$$

are equal. But (proof of Theorem 1), this is just $N_1 = N_2$.

Example. The numbers of solutions of

$$y^2 \equiv x(x^2 + ax + b) \pmod{p}$$

and

$$y^2 \equiv x(x^2 - 2ax + a^2 - 4b) \bmod p)$$

are equal, where a, b are integers and p is any prime with

$$2b(a^2 - 4b) \not\equiv 0 \pmod{p}.$$

§25. Exercises

1. Let p be prime, $p \equiv 2 \ (3)$. Show that the number of points on the elliptic curve

$$Y^2 = X^3 + B$$

defined over $\mathbf{F}_p$ is $p + 1$.
[*Hint.* Given Y, solve for X].

2. Let p be prime, $p \equiv 3 \ (4)$. Show that the number of points on the elliptic curve

$$Y^2 = X(X^2 + A)$$

defined over $\mathbf{F}_p$ is $p + 1$.
[*Hint.* Consider $\pm X$ together].

3. Let C be an elliptic curve defined over $\mathbf{F}_p$ and let $N(n)$ be the number of points defined over $\mathbf{F}_q$, where $q = p^n$. Show that there are $\alpha, \beta \in \overline{\mathbf{Q}}$ such that

$$\alpha\beta = p$$

and

$$N(n) = p^n + 1 - \alpha^n - \beta^n.$$

Hence show that all the $N(n)$ are determined by the value of $N(1)$.

Hence determine $N(2)$ for

$$Y^2 = X^3 + X + 1,$$

with $p = 3$.

[*Hint.* §24, Exercise 4].

4. [Preparation for next exercises.] Let $\mathcal{A} \supset \mathbf{Z}$ be a commutative ring without divisors of 0 [an integral domain]. Suppose that every $\lambda \in \mathcal{A}$ satisfies an equation $\lambda^2 + a\lambda + b = 0$ ($a, b \in \mathbf{Z}$, depending on λ). Show that either $\mathcal{A} = \mathbf{Z}$ or $\mathcal{A} = \mathbf{Z}[\alpha]$ for some single element $\alpha \in \mathcal{A}$.

5. Let $p \equiv 1\ (4)$ be prime and

$$C: \quad Y^2 = X(X^2 + A)$$

an elliptic curve defined over $\mathbf{F}_p$. Let $\theta \in \mathbf{F}_p$, $\theta^2 = -1$. Show that

$$\varepsilon: \quad Y \to \theta Y, \qquad X \to -X$$

is an endomorphism of C, and that $\varepsilon^2 + 1 = 0$.

Let ϕ be the Frobenius. Show that $\phi\varepsilon = \varepsilon\phi$ and deduce that

$$\phi = u + v\varepsilon$$

for some $u, v \in \mathbf{Z}$ with $u^2 + v^2 = p$.

Show, further, that the number N of points on C defined over $\mathbf{F}_p$ is

$$N = p + 1 - 2u.$$

Evaluate N for some A and p and check that u (say) $= \frac{1}{2}(p+1-N) \in \mathbf{Z}$ and satisfies $u^2 + v^2 = p$ for some $v \in \mathbf{Z}$.

6. Let $p \equiv 1\ (3)$ and let

$$C: \quad Y^2 = X^3 + B$$

be an elliptic curve defined over $\mathbf{F}_p$. Let $\theta \in \mathbf{F}_p$, $\theta^3 = 1$, $\theta \neq 1$. Show that

$$\lambda: \quad Y \to Y, \qquad X \to \theta X$$

is an endomorphism of C and that $\lambda^2 + \lambda + 1 = 0$.

Show that the Frobenius ϕ satisfies $\phi\lambda = \lambda\phi$. Now continue as in the previous Exercise.

7. Let

$$C: \quad Y^2 = X(X^2 + 4CX + 2C^2)$$

be an elliptic curve defined over $\mathbf{F}_p$, where p is prime and -2 is a quadratic residue. Show that the number N of points is of the shape

$$N = p + 1 - 2u,$$

where $u \in \mathbf{Z}$ and there is a $v \in \mathbf{Z}$ such that

$$u^2 + 2v^2 = p.$$

[*Hint.* §24, Exercise 3.]

26

Factorizing using elliptic curves

The problem of finding a factor of a given large integer has fascinated mathematicians through the ages. Recently the question has assumed practical, and indeed political, significance with the use of the products of large primes in cryptology.

It is usually (but not always) easy to prove that a given composite integer n is composite, e.g. if there is an $a > 1$ with $a^{n-1} \not\equiv 1 \bmod n$. But finding an actual nontrivial factor is a completely other matter!

For the logician, of course, the problem of factorizing an integer n is constructive. All one has to do is to test all integers $m < n^{1/2}$ for divisibility. When, say, n has 100 decimal digits, this could take longer than the age of the universe. What are needed are practical methods.

Recently H.W. Lenstra Jr. has shown that elliptic curves provide powerful methods for this problem. We will sketch one of his attacks.

Lenstra's method is suggested by Pollard's "$p-1$ method". Let n be a large integer with an unknown prime factor p. Let a be an integer and consider

$$m = \gcd(a^k - 1, n)$$

for some integer k. If $(p-1) \mid k$ then $p \mid m$. Unless we are unlucky, not all the other primes $q \mid n$ will divide m; and so m would be a nontrivial factor of n.

One does not evaluate a^k, of course, but works modulo n. There is an algorithm which works in $O(\log k)$ steps (cf. Exercises). Evaluating the gcd is cheap, using Euclid's algorithm.

Pollard's method is particularly effective if n is divisible by a prime

p for which all the prime factors of $p - 1$ are comparatively small. The accepted recipe is to take k of the shape[33]

$$k = k(b) = \prod_{q \le b} p^{e(q)},$$

where q runs through the primes and $q^{e(q)}$ is the longest power of q which is $\le b$. Here b is chosen suitably, in a way which will be described later.

The chances of success with this method of Pollard's appear to be best, when the smallest prime factor p of n is substantially smaller than $n^{1/2}$. But even, then, we may be out of luck if $p - 1$ has some largish prime factors. One can try to find a value of a whose exponent mod p is substantially smaller than $p - 1$, but that is not very promising.

Lenstra observed that Pollard's method depends on the fact that the residue classes mod p have a group structure, and that elliptic curves provide other groups which can be used for the same purpose.

Let

$$\mathcal{C}: \quad Y^2 Z = X^3 + AXZ^2 + BZ^3$$

be an elliptic curve and let (x, y, z) with $x, y, z \in \mathbf{Z}$ be a point on it. Let

$$k(x, y, z) = (x_k, y_k, z_k),$$

where $k > 1$ is an integer and $x_k, y_k, z_k \in \mathbf{Z}$. Now let p be a prime, and suppose that $\mathcal{C}$ mod p (in an obvious sense) is an elliptic curve over $\mathbf{Z}$ mod p. The mod p points form a group whose order

$$N = N_p = N_p(A, B)$$

satisfies

$$\mid N - (p + 1) \mid < 2\sqrt{p}.$$

If $N \mid k$, the point (x_k, y_k, z_k) mod p is the "point at infinity", that is

$$p \mid z_k.$$

Given A, B, x, y, z, values of (x_k, y_k, z_k) can be computed in $O(\log k)$ steps involving addition, multiplication subtraction. Since we are using homogeneous co-ordinates, there is no need to divide. The resulting values of x_k, y_k, z_k may have a common factor, but this does not disturb the conclusion that $N_p \mid k$ implies $p \mid z_k$.

Now let n be the large integer to be factorized and let $k = k(b)$ for some suitable b, as before. Then we can evaluate x_k, y_k, z_k mod n in $O(\log k)$ steps of addition, multiplication, subtraction modulo n. The

---

[33] That is, k is the lcm of the integers $\le b$.

unknown prime divisor p of n will divide z_k mod n provided that $N_p \mid k$: and then p divides

$$m = \gcd(n, z_k).$$

If $z_k = 0 \pmod{n}$, we are out of luck. Otherwise m will be a nontrivial divisor of n: which is what we want.

It can, of course, happen that $m = 1$, if $N_p \nmid k$ for all $p \mid n$. If this happens, we select other values of A, B, x, y, z (and, possibly, k) and try, try, try again.[34]

The above account leaves a couple of questions unanswered.

(i) How do we choose the initial curve C and the point (x, y, z)? Since all the calculations are mod n, it is enough to find A, B, x, y, $z \in \mathbf{Z}$ such that

$$y^2 z \equiv x^3 + Axz^2 + Bz^3 \qquad \text{mod } n.$$

An obvious way is to put $z = 1$, choose A, x, y at random and use the equation to determine B. Since we naturally suppose that we started off by checking that n has no small divisors, the chance that C is not an elliptic curve for any $p \mid n$ is negligible. In any case, there is no harm in running through the algorithm: at worst we will draw a blank. Alternatively, one can compute

$$l = \gcd(n, 4A^3 + 27B^2).$$

If $l = 1$, we are OK. If $1 < l < n$, we have a non-trivial factor of n, which is what we want. If $l = n$, which is highly unlikely, we abort the run and choose fresh A, B, x, y, z.

(ii) What is the optimal choice of b in $k = k(b)$? It turns out that this depends on the smallest prime divisor p of n: which is, of course, unknown. We argue heuristically.

Let $1 < s < t$, where t is an integer. We say that t is s-smooth if every prime divisor q of t is less than s. It is known that the number of integers $t < T$, for given T, which are $T^{1/u}$-smooth is very roughly $u^{-1/u}T$. Put

$$L = L(T) = \exp(\sqrt{(\log T \log \log T)})$$

and let $0 < \alpha < \infty$. On putting $T^{1/u} = L^\alpha$, we deduce that the number of $t \leq T$ which are L^α-smooth is roughly $L^{-1/2\alpha}T$. We shall paraphrase

---

[34] For the distribution of N_p over curves, see B.J. Birch: How the number of points of an elliptic curve over a fixed prime varies. *J. London Math. Soc.* **43** (1968), 57-60.

this to the statement that the probability P that a random integer t in the neighbourhood of T is L^α-smooth is $P = L^{-1/2\alpha}$.

We shall choose the best value of α later. Let p be the unknown smallest prime factor of n. Put $L = L(p)$. We have seen that the order $N_p = N_p(A, B)$ of the points mod p on $\mathcal{C}$ is approximately p. Assuming that N_p behaves reasonably randomly as A, B vary, the probability P that N_p is L^α-smooth is $P = L^{-1/2\alpha}$.

Take

$$b = L^\alpha, \qquad k = k(b).$$

Then all the prime factors of N_p divide k. The practitioners of the mystery of factorization assume that it is highly probable that indeed $N_p \mid k$, which we suppose. The number of steps in one run of the algorithm is $O(\log k)$, which is very roughly $b = L^\alpha$.

To sum up. The amount of work in a run of the algorithm is about L^α. The probability of success in a single run is about $L^{-1/2\alpha}$. Hence the expected work to find a nontrivial factor is about

$$L^{\alpha+1/2\alpha}.$$

This is minimized at

$$\alpha = 1/\sqrt{2},$$

which is therefore the optimal choice.

The above estimates depend on the size of the unknown least prime factor p of n. The worst case scenario is when p is nearly $n^{1/2}$. However, one expects the Lenstra algorithm to be most effective when the smallest prime factor is much smaller. Thus it works better on "naturally occurring" integers n than on the integers n used in some cryptosystems, which are the product of two nearly equal primes. If nothing is known a priori about the primes in n, a good strategy is to start with a comparatively small b and to increase it gradually if necessary.

We have chosen a version of the Lenstra algorithm which is easy to describe, rather than one which minimizes computation time. In practice, further devices and stratagems are brought into play. We do not go into this here, but conclude with a variant in the spirit of the course.

In the variant, one considers the elliptic curve

$$\mathcal{C}: \quad CY^2 = X^3 + AX + B$$

for some $C \neq 0$, where we now take the inhomogeneous form. Recall that if (x_1, y_1) and (x_2, y_2) are points on the curve and

$$(x_3, y_3) = (x_1, y_1) + (x_2, y_2),$$

$$(x_4, y_4) = (x_1, y_1) - (x_2, y_2)$$

then x_3, x_4 are the roots of a quadratic equation whose coefficients are polynomial in x_1, x_2, A, B (but not C). If now k is a positive integer and if, to change the notation, (x, y) is a rational point on C and $(x_k, y_k) = k(x, y)$, then the classical algorithm for computing g^k can be modified to give an algorithm to compute x_k in $O(\log k)$ steps (cf. Exercises).

Now write $X = U/V$ and work homogeneously. If $x = u/v$, then $x_k = x_k/v_k$ where u_k, v_k are obtained from u, v by $O(\log k)$ additions, subtractions and multiplications, but no divisions.

Now, as before, let n be the number to be factorized and p an unknown prime divisor. Suppose that A, B, u, $v \in \mathbb{Z}$ and work mod n. then, as before, if $N_p \mid k$ then $p \mid v_k$ and we can expect that $\gcd(v_k, n)$ is a non-trivial divisor of n.

In this version of the algorithm we may choose A, B, u, v entirely arbitrarily. Put $x = u/v$, $y = 1$. Then, unless we are strikingly unlucky, the point (x, y) lies on C for some $C \in \mathbb{Q}^*$ which need not be evaluated, as it is never needed.

Elliptic curves are used also in primality testing and in other unexpected ways: for example, finding square roots modulo a large prime. See A.K. Lenstra and H.W. Lenstra Jr., Algorithms in number theory. Chapter 12 (pp.673-715) of: *Handbook of theoretical computer science*, vol. A (ed. J. van Leeuwen), Elsevier, 1990.

§26. Exercises

1. [Motivation for next question.] Let G be an abelian group and n a positive integer. For $g \in G$ show that the following algorithm computes g^n in $O(\log n)$ operations.

(i) $N = n$, $Y = 1 \in G$, $Z = g$
(ii) IF $N = 0$, GOTO END
(iii) $M = [N/2]$, $E = N - 2M$
(iv) IF $E = 1$ THEN $Y = YZ$
(v) $N = M$, $Z = Z^2$
(vi) GOTO (ii)
      END $[Y = g^n]$.

2. Let $C : \quad Y^2 = X^3 + AX + B$ be an elliptic curve. For positive *odd* integer n and $\mathbf{a} = (a, b)$ on C, check that the following algorithm computes u, where $n\mathbf{a} = (u, v)$, in $O(\log n)$ steps.

We recall that there is a rational function $d(x)$ such that if $\mathbf{x} = (x, y)$ then $2\mathbf{x} = (d(x), ?)$ for some $?$.

Further, there is a quadratic $q(T) = q(T; x_1, x_2)$ whose coefficients are rational in x_1, x_2 and whose roots are x_3, x_4 if $\mathbf{x_3} = \mathbf{x_1} + \mathbf{x_2}$, $\mathbf{x_4} = \mathbf{x_1} - \mathbf{x_2}$.

(i) $N = n$, $X = a$, $Y = a$, $Z = a$

(ii)   IF $N = 0$, GOTO END

(iii) $M = [N/2]$, $E = N - 2M$.

(iv) $Z = d(Z)$.

(v)   IF $E = 1$ GOTO (viii)

(vi) [Check that Y is a root of $q(T; X, Z)$.] Y IS THE OTHER ROOT OF $q(T; X, Z)$

(vii) GOTO (ix)

(viii) [Check that X is a root of $Q(T; Y, Z)$.] X IS THE OTHER ROOT OF $q(T; Y, Z)$.

(ix) $N = M$

(x)   GOTO (ii).

     END $[X = u$, where $n(a, b) = (u, v)$.$]$

3. Suppose that (a, b) lies on

$$C^* : \quad EY^2 = X^3 + AX + B$$

for some $E \neq 0$. Let $n(a, b) = (u, v)$ on C^*. Show that u is given by the algorithm in (2). [i.e. the algorithm is independent of E.]

Formulary

Desboves' Formulae[35]. These are for

$$a_1 X_1^3 + a_2 X_2^3 + a_3 X_3^3 + d X_1 X_2 X_3 = 0.$$

This is nonsingular if $27 a_1 a_2 a_3 + d^3 \neq 0$. The residual intersection **t** of the tangent at **x** is

$$t_j = x_j(a_{j+1} x_{j+1}^3 - a_{j+2} x_{j+2}^3) \qquad (j \text{ taken mod } 3)$$

The third intersection **z** of the line joining **x**, **y** is

$$z_j = x_j^2 y_{j+1} y_{j+2} - y_j^2 x_{j+1} x_{j+2} \qquad (j \bmod 3).$$

Canonical curve.

$$Y^2 = X^3 + AX + B.$$

If $\mathbf{x} = (x, y)$, then $-\mathbf{x} = (x, -y)$.

Addition formula. Let

$$\mathbf{x}_1 = (x_1, y_1); \qquad \mathbf{x}_2 = (x_2, y_2).$$

and

$$\mathbf{x} = \mathbf{x}_1 + \mathbf{x}_2, \qquad \mathbf{x} = (x, y).$$

---

[35] A. Desboves. Résolution en nombres entiers et sous sa forme la plus générale, de l'équation cubique, homogène à trois inconnues. *Nouv. Ann. de la Math.*, Sér. III, vol. 5 (1886), 545-579.

If $x_2 = -x_1$, we have $x = 0$. If $x_2 = x_1$, we apply the duplication formula, given below. Otherwise, we may suppose that

$$x_2 \neq x_1.$$

The line joining x_1, x_2 is

$$Y = lX + m,$$

where

$$l = \frac{y_1 - y_2}{x_1 - x_2}, \qquad m = \frac{x_1 y_2 - x_2 y_1}{x_1 - x_2}.$$

This line cuts the curve in x_1, x_2 and

$$-(x_1 + x_2) = -x = (x, -y).$$

The roots of

$$X^3 + AX + B - (lX + m)^2$$

are x_1, x_2 and x. Hence

$$x = l^2 - x_1 - x_2;$$

and so

$$(x_1 - x_2)^2 x = x_1 x_2^2 + x_1^2 x_2 - 2y_1 y_2 + A(x_1 + x_2) + 2B.$$

Further,

$$y = -lx - m;$$

and so

$$(x_1 - x_2)^3 y = W_2 y_2 - W_1 y_1,$$

where

$$W_1 = 3x_1 x_2^2 + x_2^3 + A(x_1 + 3x_2) + 4B$$
$$W_2 = \text{symmetric.}$$

Duplication formula. Here we consider

$$(x_2, y_2) = x_2 = 2x = 2(x, y).$$

If $y = 0$ we have $x_2 = 0$. Hence we may suppose

$$y \neq 0.$$

We need the tangent

$$Y = lX + m$$

at x. Since formal differentiation on the curve gives

$$2Y \frac{dY}{dX} = 3X^2 + A,$$

we have

$$l = (3x^2 + A)/2y.$$

Hence (as for addition formula)
$$x_2 = l^2 - 2x$$
$$= \frac{(3x^2 + A)^2 - 8xy^2}{4y^2};$$

i.e.
$$x_2 = \frac{x^4 - 2Ax^2 - 8Bx + A^2}{4(x^3 + Ax + B)}.$$

To find y_2 we need the value
$$m = \frac{-x^3 + Ax + 2B}{2y},$$
which is determined by $y = lx + m$. Now
$$y_2 = -lx_2 - m;$$
which gives
$$(2y)^3 y_2 = x^6 + 5Ax^4 + 20Bx^3$$
$$- 5A^2x^2 - 4ABx - A^3 - 8B^2.$$

Formulae in X only. Let
$$\mathbf{x}_1 = (x_1, y_1), \qquad \mathbf{x}_2 = (x_2, y_2)$$
with
$$x_1 \neq x_2.$$

Let
$$\mathbf{x}_3 = \mathbf{x}_1 + \mathbf{x}_2 = (x_3, y_3)$$
$$\mathbf{x}_4 = \mathbf{x}_1 - \mathbf{x}_2 = (x_4, y_4).$$
Then
$$(x_1 - x_2)^2(x_3 + x_4) = 2(x_1x_2 + A)(x_1 + x_2) + 4B,$$
$$(x_1 - x_2)^2 x_3 x_4 = x_1^2 x_2^2 - 2Ax_1x_2 - 4B(x_1 + x_2) + A^2.$$

This follows from the expression for x in the addition formula. The value of x_3 is x as given and that of x_4 is obtained from it merely by changing the sign of y_1y_2. Hence the formula for $x_3 + x_4$ is immediate. That for x_3x_4 comes by substituting for $y_1^2 y_2^2$ in the product and cancelling $(x_1 - x_2)^2$. [Alternatively, cf. §17, Exercise 3.]

Multiplication[36]. Let $(X_m, Y_m) = m(X, Y)$ where $m \in \mathbf{Z}$. Then

$$X_M = \frac{X\psi_m^2 - \psi_{m-1}\psi_{m+1}}{\psi_m^2},$$

$$Y_m = \frac{\psi_{2m}}{2\psi_m^4},$$

where

$\psi_0 = 0,$

$\psi_1 = 1,$

$\psi_2 = 2Y,$

$\psi_3 = 3X^4 + 6AX^2 + 12BX - A^2,$

$\psi_4 = 4Y(X^6 + 5AX^4 + 20BX^3 - 5A^2X^2 - 4ABX - 8B^2 - A^3),$

$\psi_{2n+1} = \psi_n^3\psi_{n+2} - \psi_{n+1}^3\psi_{n-1},$

$Y\psi_{2n} = \psi_n\{\psi_{n-1}^2\psi_{n+2} - \psi_{n+1}^2\psi_{n-2}\}.$

This is an exercise on the fact that a function is defined up to multiplicative constant by its zero and poles. We determine the constants by looking at the behaviour at o using the local uniformiser

$$t = X/Y.$$

ψ_m is defined by

(i)   it has a simple zero at all a $\neq$ o with ma $= 0$. (a defined over $\overline{\mathbf{Q}}$).

(ii)   it behaves like mt^{-m^2+1} at o.

More precisely

(I)   if m is odd, there are $\frac{1}{2}(m^2 - 1)$ pairs $(a_j, \pm b_j)$ of m-division pairs and

$$\psi_m = m\prod(X - a_j).$$

(II)   If m is even, the three 2-division points are m-division points, and there are $\frac{1}{2}(m^2 - 4)$ pairs $(a_j, \pm b_j)$, $b_j \neq 0$. Then

$$\psi_m = mY\prod(X - a_j).$$

Now for all m, even or odd, we have

$$X_m \sim m^{-2}t^{-2}, \qquad Y_m \sim m^{-3}t^{-3}$$

---

[36] cf. H. Weber, *Algebra III*, §58; but we have adjusted the sign of ψ_m so that the leading term is always positive.

at o, and

$$\psi_m^2 X_m$$

has no poles except at o.

Further, $X_m - X$ vanishes at a only if $(m+1)\mathbf{a} = \mathbf{o}$ or $(m-1)\mathbf{a} = \mathbf{o}$. Hence

$$X - X_m = \psi_{m+1}\psi_{m-1}/\psi_m^2, \qquad (*)$$

where the constant is right since both sides behave like $(m^2 - 1)/m^2 t^2$ at o. This gives the formula for X_m. That for Y_m follows immediately from the specification of the poles and zeros.

It remains to give the recurrence relation. For integers l, m we have $X_l = X_m$ precisely when either $(l+m)(X, Y) = \mathbf{o}$ or $(l-m)(X, Y) = \mathbf{o}$. Hence

$$X_l - X_m = \psi_{m+l}\psi_{m-l}/\psi_l^2\psi_m^2;$$

the constant being determined by the behaviour at o. But

$$X_l - X_m = (X - X_m) - (X - X_l)$$

Hence by (*)

$$\psi_l^2\psi_{m+1}\psi_{m-1} - \psi_m^2\psi_{l+1}\psi_{l-1}$$
$$= \psi_{m+l}\psi_{m-l}.$$

Put $l = n$, $m = n+1$, so $\psi_{m-l} = 1$ and

$$\psi_{2n+1} = \psi_n^3\psi_{n+2} - \psi_{n+1}^3\psi_{n-1}$$

Put $l = n-1$, $m = n+1$ so $\psi_{m-l} = \psi_2 = Y$. Then

$$Y\psi_{2n} = \psi_n\{\psi_{n-1}^2\psi_{n+2} - \psi_{n+1}^2\psi_{n-1}\}.$$

Further Reading

Cassels, J.W.S. Diophantine equations with special reference to elliptic curves, *J. London Math. Soc.* **41** (1966), 193-291.

Husemöller, D. *Elliptic curves*, Springer, 1987.

Koblitz, N. *Introduction to elliptic curves and modular forms*, Springer, 1984.

Lang, S. *Fundamentals of diophantine geometry*, Springer, 1983. [The first edition is less complete but more coherent: *Diophantine geometry*, Interscience, 1962.]

Serre, J.-P. *Lectures on the Mordell-Weil theorem*, Vieweg, 1989. [Notes of a course given in 1980-81]

Silverman, J.H. *The arithmetic of elliptic curves*, Springer, 1986.

Tate, J. The arithmetic of elliptic curves, *Invent. Math.* **23** (1974), 179-206.

INDEX

multiplicity   23, 44

Nagell   34(fn), 52(fn)
neutral element (of group)   27
Newton   24, 43
nonsense   98 *et seq*
non-archimedean   7
non-singular   24
norm (map)   66

patch   67
pole   30
Pollard   124
principal homogeneous spaces   104
        et seq
p-adic filtration   48
p-adic integers   9
p-adic numbers   6
p-adic units   9
p-adic valuation   7

rational curve
        (= curve of genus 0)   3
rational (point etc.)   3
reducible (curve):
        see also irreducible   43(fn)
reduction mod p 42 *et seq*
Reichardt   85
resultant   75 *et seq*
"Riemann hypothesis for
        function fields"   2, 119
Riemann-Roch theorem   30
Rubin   111

Schmidt   120
Selmer   87, 110
Shafarevich   85
singular (point)   23
Swinnerton-Dyer   71, 110
symmetric (pointset)   18

Tamagawa number   110
Tate   85, 109(fn)
Tate-Shafarevich group   85, 109 *et
        seq*
torsion   102
triangle inequality   7

ultrametric inequality   7
unit (p-adic)   9

valuation   6
valuation (p-adic)   7
van der Corput   19

weak finite basis
        theorem   55, 66 *et seq*
Weil   1, 54, 108, 119
Weil-Châtelet group   108

Printed in the United States
By Bookmasters